Vibration Analysis with SOLIDWORKS Simulation 2019

Paul M. Kurowski, Ph.D., P.Eng.

CERTIFIED
Solution
Partner

DS SOLIDWORKS

SDC
Publications

Design Generator, Inc.

SDC Publications
P.O. Box 1334
Mission, KS 66222
913-262-2664
www.SDCpublications.com
Publisher: Stephen Schroff

All models used in this book may be downloaded from www.SDCPublications.com.

ISBN-13: 978-1-63057-243-3
ISBN-10: 1-63057-243-8

Printed and bound in the United States of America.

About the cover

The image on the cover shows a higher order mode of vibration of MESH2019 part.

The model, complete with SOLIDWORKS Simulation study, is located in the folder *00 Cover Page* in the set of exercises that accompany this book.

SOLIDWORKS Models

All models used in this book may be downloaded from www.SDCPublications.com

About the Author

Dr. Paul Kurowski obtained his MSc and PhD in Applied Mechanics from Warsaw Technical University. He completed postdoctoral work at Kyoto University. Dr. Kurowski is an Assistant Professor in the Department of Mechanical and Materials Engineering at Western University. His teaching includes Finite Element Analysis, Product Design, Kinematics and Dynamics of Machines and Mechanical Vibrations. His interests focus on Computer Aided Engineering methods used as tools of product design.

Dr. Kurowski is also the President of Design Generator Inc., a consulting firm with expertise in Product Development, Design Analysis, and training in Computer Aided Engineering.

Dr. Kurowski has published many technical papers and taught professional development seminars for the SAE International, the American Society of Mechanical Engineers (ASME), the Association of Professional Engineers of Ontario (PEO), the Parametric Technology Corporation (PTC), Rand Worldwide, SOLIDWORKS Corporation and others.

Dr. Kurowski is a member of the Association of Professional Engineers of Ontario and the SAE International. He can be contacted at www.designgenerator.com.

Acknowledgements

I would like to thank the students attending my various courses for their valuable comments and questions. I thank my wife Elżbieta for editing, proofreading and project management.

Paul Kurowski

Table of contents

Before you start   **1**

Notes on hands-on exercises and functionality of Simulation
Prerequisites
Selected terminology

1: Introduction to vibration analysis   **5**

Differences between a mechanism and a structure
Difference between dynamic analysis and vibration analysis
Rigid body motion and degrees of freedom
Kinematic pairs
Discrete and distributed vibration systems
Single degree of freedom and multi degree of freedom vibration systems
Mode of vibration
Rigid Body Mode
Modal superposition method
Direct integration method
Vibration Analysis with SOLIDWORKS Simulation and SOLIDWORKS Motion
Functionality of SOLIDWORKS Simulation and SOLIDWORKS Motion
Terminology issues

2: Introduction to modal analysis   **35**

Modal analysis
Properties of a mode of vibration
Interpreting results of modal analysis
Normalizing displacement results in modal analysis

3: Modal analysis of distributed systems   **45**

Modal analysis of distributed systems
Meshing considerations in modal analysis
Importance of mesh quality in modal analysis
Importance of modeling supports
Interpretation of results of modal analysis

Prerequisites

If you are new to **SOLIDWORKS Simulation** please read "**Engineering Analysis with SOLIDWORKS Simulation 2019**" to gain essential familiarity with Finite Element Analysis as implemented in **SOLIDWORKS Simulation**. Complete exercises in chapters 4, 6, 18, 19 of that textbook.

The following prerequisites are recommended:

- An understanding of Vibration Analysis
- An understanding of Structural Analysis
- An understanding of Solid Mechanics
- Familiarity with **SOLIDWORKS**
- Familiarity with **SOLIDWORKS Simulation** to the extent covered in "**Engineering Analysis with SOLIDWORKS Simulation**" or an equivalent experience
- Familiarity with Windows OS

Selected terminology

The mouse pointer plays a very important role in executing various commands and providing user feedback. The mouse pointer is used to execute commands, select geometry, and invoke pop-up menus. We use Windows terminology when referring to mouse-pointer actions.

Item	Description
Click	Self-explanatory
Double-click	Self-explanatory
Click-inside	Click the left mouse button. Wait a second, and then click the left mouse button inside the pop-up menu or text box. Use this technique to modify the names of folders and icons in **SOLIDWORKS Simulation** Manager.
Drag and drop	Use the mouse to point to an object. Press and hold the left mouse button down. Move the mouse pointer to a new location. Release the left mouse button.
Right-click	Click the right mouse button. A pop-up menu is displayed. Use the left mouse button to select a desired menu command.

All **SOLIDWORKS** file names appear in CAPITAL letters, even though the actual file names may use a combination of capital and small letters. Selected menu items and **SOLIDWORKS Simulation** commands appear in **bold**; **SOLIDWORKS** configurations, **SOLIDWORKS Simulation** folders, icon names and study names appear in *italics* except in captions and comments to illustrations. **SOLIDWORKS** and **Simulation** also appear in bold font. Bold font is also used to draw the reader's attention to a particular term.

A **planar** kinematic pair (PLANAR.SLDASM) is shown in Figure 1-5.

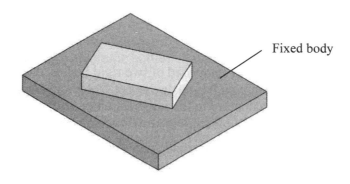

Fixed body

Figure 1-5: A planar kinematic pair removes three degrees of freedom from the moving body.

Two linear coordinates are required to define the position of the center of the sliding block and one angular coordinate is required to define its angular position. Therefore, the pair has two translational degrees of freedom and one rotational degree of freedom.

A **prismatic** kinematic pair (PRISMATIC.SLDASM) is shown in Figure 1-6.

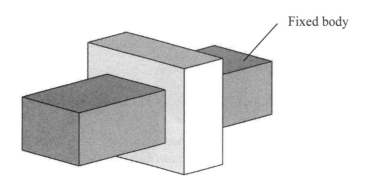

Fixed body

Figure 1-6: A prismatic kinematic pair removes five degrees of freedom from the moving body.

One linear coordinate is required to define the position of the sliding block. This pair has one translational degree of freedom.

Notice that the "Deformed shape" shown in Figure 1-14 refers to the deformation of the spring even though the spring deformation can't be seen on this illustration. The cylinder itself is not deforming, it performs linear oscillations as a rigid body. All this applies to the first mode of vibration of this assembly.

The SWING ARM model also has a **Frequency** study defined; please run the solution before proceeding.

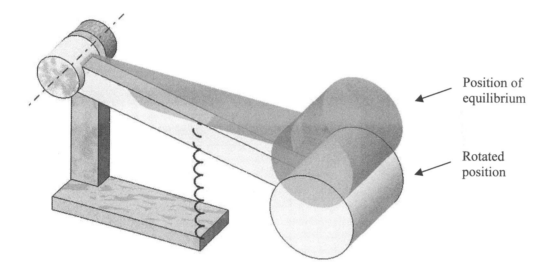

Figure 1-15: Shape of vibration of SWING ARM model.

The arm rotates as a rigid body; it does not experience any deformation. The only deforming element is the spring. Motion of the arm is controlled by a hinge modeled as a Fixed Hinge restraint.

Review assembly model SWING ARM and notice that the vertical post in the base serves only to locate the hinge position of the swing arm. The hinge itself is not modeled. The hinge is simulated in the **Frequency** study using a **Fixed Hinge** restraint.

A **revolute** kinematic pair (REVOLUTE.SLDASM) is shown in Figure 1-7.

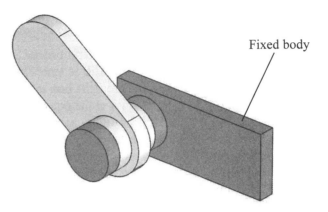

Figure 1-7: A revolute kinematic pair removes five degrees of freedom from the moving body.

One angular coordinate is required to define the position of the moving part. The pair has one rotational degree of freedom.

A **cylindrical** kinematic pair (CYLINDRICAL.SLDASM) is shown in Figure 1-8.

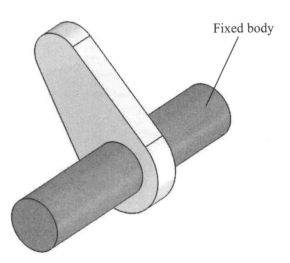

Figure 1-8: A cylindrical kinematic pair removes four degrees of freedom from the moving body.

One linear coordinate and one angular coordinate are required to define the position of the moving part. The pair has one translational and one rotational degree of freedom.

Mode of vibration

The mode of vibration is the fundamental property of a vibrating system. Its central role in vibration analysis requires us to discuss it in this introductory chapter. We will refine this definition later.

The mode of vibration can be defined as the preferred way of a structure to vibrate. It is characterized by its frequency of vibration, mass participating in the vibration, and shape of vibration. The number of modes of vibration is equal to the number of degrees of freedom of the vibrating system.

Shapes of vibration in the first (and only) mode of the discrete vibration system DISCRETE LINEAR in the *1DOF* configuration is shown in Figure 1-14. This model is available with **Frequency** studies set up and ready to run. Please run the solutions before proceeding.

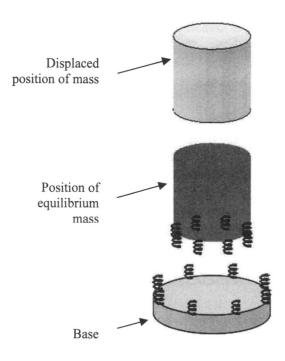

Displaced position of mass

Position of equilibrium mass

Base

<u>Figure 1-14: The shape of vibration in the first mode of the DISCRETE LINEAR model. The moving mass translates as a rigid body; it does not experience any deformation. The only deforming element is the spring.</u>

The mass moves as a rigid body; its motion is restricted to translation by restraints defined in SOLIDWORKS Simulation. Two assembly components are connected by a distributed spring connector also defined in SOLIDWORKS Simulation.

A two degree of freedom discrete system has two modes of vibration, each one characterized by its own unique shape as demonstrated by the DOUBLE PENDULUM model, shown in Figure 1-16. Run the **Frequency** study before proceeding.

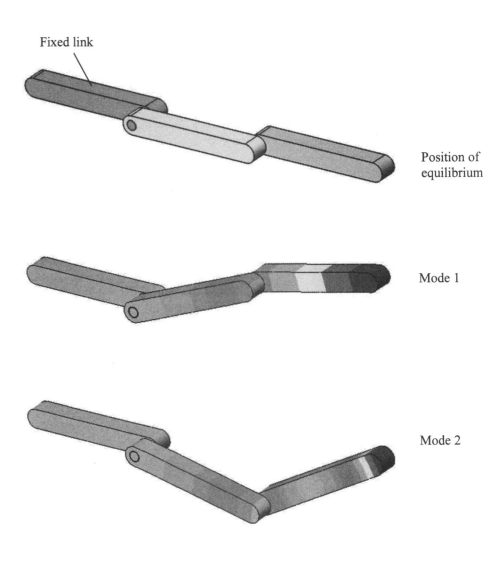

Fixed link

Position of equilibrium

Mode 1

Mode 2

<u>Figure 1-16: Shapes of vibration of DOUBLE PENDULUM model.</u>

Plots of Mode 1 and Mode 2 use an arbitrary scale of deformation.

Review the **Frequency** study in the DOUBLE PENDULUM model. Notice that **Global Contact** is set to **Allow Penetration** in order to disconnect the touching faces which otherwise would be bonded. The three links are connected by two **Pin Connectors** with no torsional stiffness.

In distributed systems the deformation is not limited to discrete springs. These systems have an infinite number of degrees of freedom and consequently, an infinite number of modes of vibration. Consider CLIP (Figure 1-13) to be unsupported. CLIP treated as a rigid body has six degrees of freedom: three translations and three rotations; we may say it has six rigid body motions. When treated as an elastic body, CLIP will have an infinite number of degrees of freedom associated with elastic deformation in addition to those six degrees of freedom associated with **Rigid Body Motions (RBMs)**. Consequently, it has an infinite number of modes of vibration. When this CLIP is analyzed using FEA, the number of elastic degrees of freedom becomes finite due to discretization which is an inherent part of FEA (Figure 1-17).

CAD model

Finite element mesh

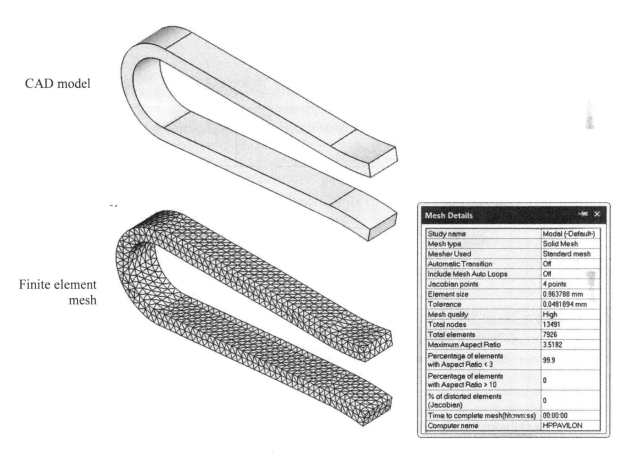

Mesh details

Figure 1-17: CLIP model before discretization (top) and after discretization (bottom).

Meshing was performed using second order tetrahedral solid elements with a default element size. This produced a mesh with 13491 nodes as shown in Mesh Details window.

The principle of the **Modal Superposition Method** is shown schematically in Figure 1-21.

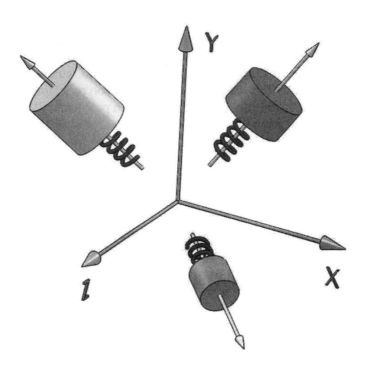

Figure 1-21: A vibration system represented by three single degree of freedom oscillators.

It is assumed that three modes of vibration are sufficient to model the vibration response. Each mode is represented by a SDOF oscillator. Each oscillator is characterized by its mass, stiffness and direction of linear displacement.

The **Modal Superposition Method** is universally used in vibration analysis with FEA. The most important condition that must be satisfied in order to use the **Modal Superposition Method** is that the problem must be **linear**. The second important issue is how many modes should be considered? This is most often decided by considering the frequency range of excitation. A convergence process may also be conducted where the same problem is solved a number of times, each time with more modes calculated to find the sensitivity of the data of interest to the number of modes considered in the **Modal Superposition Method**.

Direct Integration Method

An alternative to the **Modal Superposition Method** is the **Direct Integration Method**. It works with all degrees of freedom of the model. It does not require an assumption of the number of modes to be considered in a vibration response and the problem doesn't have to be linear. The **Direct Integration Method** is available in **SOLIDWORKS Simulation** but its numerical intensity limits practical applications to vibration problems of short time duration.

Vibration Analysis with SOLIDWORKS Motion and SOLIDWORKS Simulation

We have already defined discrete and distributed vibration systems; these definitions are essential to understanding differences between **SOLIDWORKS Simulation** and **SOLIDWORKS Motion** which are add-ins to **SOLIDWORKS**.

Vibration of discrete systems can be analyzed both by **SOLIDWORKS Motion** and **SOLIDWORKS Simulation**. Vibration of distributed systems can be analyzed only with SOLIDWORKS **Simulation.**

SOLIDWORKS Motion is a tool for kinematic and dynamic analyses of rigid bodies. Elasticity can be modeled in **SOLIDWORKS Motion** only as a discrete spring which makes it possible to use it for vibration analysis of discrete systems. Models analyzed by **SOLIDWORKS Motion** are typically mechanisms and have few degrees of freedom associated with rigid body motions of its components. All deformation is limited to springs. Vibration properties of distributed systems cannot be analyzed with **SOLIDWORKS Motion**.

SOLIDWORKS Simulation is a tool of structural analysis of elastic bodies by means of Finite Element Analysis. **SOLIDWORKS Simulation** models everything as elastic bodies with a large number of degrees of freedom and, consequently, with a large number of modes of vibration. It is up to the user to decide how many modes should be used to model a vibration response. Using **SOLIDWORKS Simulation**, a discrete representation of a problem does not offer any advantages besides a more intuitive understanding of the problem and the results. **SOLIDWORKS Simulation** performs discretization and represents every problem as a discrete system with a large but finite number of degrees of freedom.

Figure 1-22 shows the ELLIPTIC TRAMMEL model. This device can be used to trace an ellipse, hence the name. **SOLIDWORKS Motion** treats it as a mechanism. This mechanism has one degree of freedom and consists of two prismatic kinematic pairs and two revolute kinematic pairs. The angular position of the tracer or linear position of either slider fully defines the position of this mechanism. The elliptic trammel may become a vibration system if a torsional spring is added to either hinge. Then, any displacement will be associated with deformation, and the system will satisfy the requirements to be classified as a structure. Notice that the model with the torsional spring added still has one degree of freedom.

If the same model (with or without a spring) is analyzed with **SOLIDWORKS Simulation**, all components are treated as elastic bodies. The number of degrees of freedom will depend on the element size. If the default element size is used, the meshed model will have 14375 nodes and 43125 degrees of freedom.

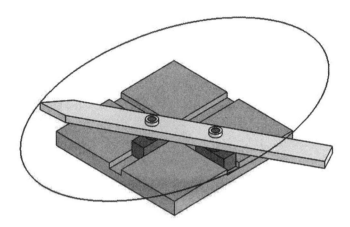

The model analyzed by **SolidWorks Motion** has one degree of freedom. One piece of information fully defines the position of the mechanism.

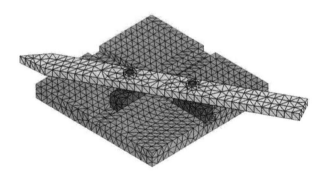

This model analyzed by **SolidWorks Simulation** has 14260 nodes; it translates into over 42000 degrees of freedom; this number will change depending on the element size and definition of restraints.

Figure 1-22: Different representations of the ELLIPTIC TRAMMEL model by SOLIDWORKS Motion (top) and SOLIDWORKS Simulation (bottom).

SOLIDWORKS Motion model is shown with the trajectory of motion of the tip of link; this is an ellipse. Animate the model in SOLIDWORKS motion and review mesh details in SOLIDWORKS Simulation.

SOLIDWORKS Simulation and **SOLIDWORKS Motion** are connected to **SOLIDWORKS** CAD as well as between themselves. Connectivity and analysis capabilities are shown in Figure 1-25.

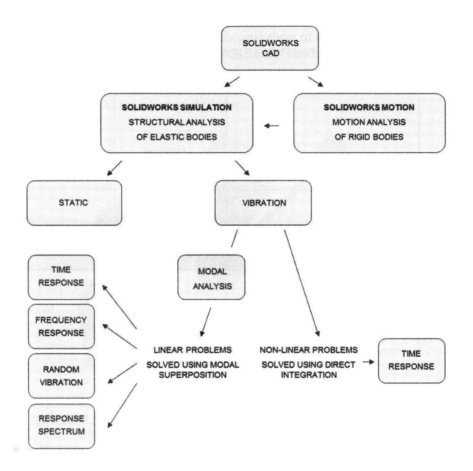

Figure 1-25: Types of analyses available in SOLIDWORKS Simulation and connectivity between SOLIDWORKS Simulation and SOLIDWORKS Motion.

Results of SOLIDWORKS Motion may be transferred to SOLIDWORKS Simulation.

SOLIDWORKS Motion will not be used in this book. Most vibration problems in this book will be solved with **SOLIDWORKS Simulation** using the **Modal Superposition Method**. These will be problems of **Time Response**, **Frequency Response**, **Random Vibration** and **Response Spectrum**. Nonlinear vibration problems will be solved using the **Direct Integration Method**. We will provide in-depth descriptions of these types of analyses later, but now we have to address some important terminology issues.

Terminology issues

Terms used in **SOLIDWORKS Simulation** differ in many ways from terms you find in a standard textbook on vibration analysis. A summary of important differences along with short descriptions are given in Figure 1-26.

Textbook terminology	SOLIDWORKS Simulation study name	Description
Linear Vibration Analysis	Linear Dynamic	All types of linear vibration analyses based on the Modal Superposition Method
Nonlinear Vibration Analysis	Nonlinear with Dynamic Option	Nonlinear vibration analysis; Excitation is a function of time
Modal	Frequency	Finds modes of vibration (frequency and shape)
Time Response	Modal Time History in linear analysis Nonlinear with Dynamic Option in nonlinear analysis	Excitation is a function of time
Frequency Response	Harmonic	Excitation is harmonic; defined as a function of frequency
Random Vibration	Random	Excitation is in the form of a Power Spectral Density
Response Spectrum	Response Spectrum Analysis	Base excitation in the form of a Response Spectrum
	Drop test	Nonlinear vibration analysis intended for drop test simulation

Figure 1-26: Summary of capabilities of SOLIDWORKS Simulation and SOLIDWORKS Motion.

We will be alternating between standard textbook terminology and SOLIDWORKS Simulation terminology.

2: Introduction to modal analysis

Topics covered

- Modal analysis
- Properties of a mode of vibration
- Interpreting results of modal analysis
- Normalizing displacement results in modal analysis

Modal analysis

Modal analysis, called a **Frequency** analysis in **Simulation**, finds natural frequencies and shapes of vibration that are associated with natural frequencies. **Modal** analysis assumes that the body vibrates in the absence of any excitation and damping. The vibration textbooks call this free undamped vibration.

Consider the simplest type of vibrating system: a **Single Degree of Freedom** oscillator (SDOF) with mass m and stiffness k which is vibrating in the absence of damping, excitation and gravity (Figure 2-1).

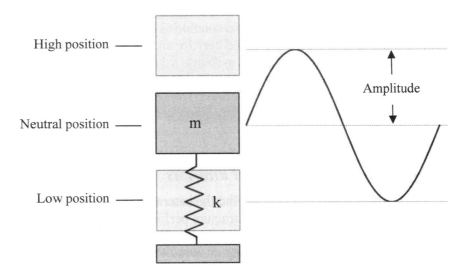

Figure 2-1: SDOF oscillator vibrating in the absence of damping and excitation. The problem is linear; mass m and stiffness k do not change during motion.

Vibration has been caused by an initial condition such as a displacement and/or a velocity.

Numerical values of displacement results are normalized to make the modal mass equal to 1. Therefore, the displacement results will be different for the same model geometry and stiffness but different mass. To demonstrate this, we modify the original density of 7800kg/m³ and repeat the analysis with two custom materials:

1. Custom material 1: modulus of elasticity equal to that of Cast Carbon steel; mass density 780kg/m³

2. Custom material 2: modulus of elasticity equal to that of Cast Carbon steel; mass density of 78000kg/m³

Copy the frequency study *01 Cast Carbon Steel* into two new studies titled *02 10% Cast Carbon Steel* and *03 1000% Cast Carbon Steel*. Assign the corresponding custom materials as explained above. Obtain the results and summarize the displacements as shown in Figure 2-5.

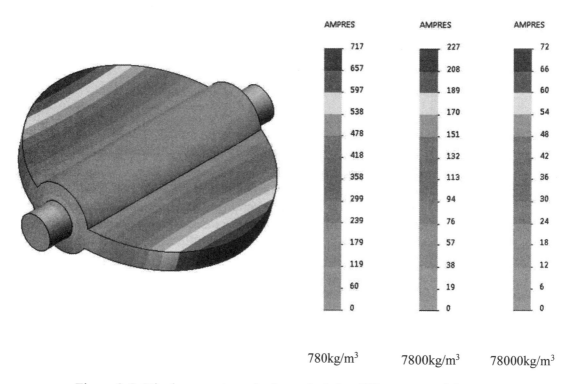

780kg/m³   7800kg/m³   78000kg/m³

Figure 2-5: Displacement results in mode 1 for different materials.

Color legends are shown in the ascending order or mass density. The modal shape is the same for all material densities.

To gain a better insight on how displacement results are normalized in Modal analysis, we'll summarize the natural frequencies and the maximum displacement magnitudes for the different material densities listed in Figure 2-5. The summary is shown in Figure 2-6.

	Material mass density kg/m^3	Frequency of the first mode	Normalized max. displacement
1	780	1166Hz	717
2	7800	370Hz	227
3	78000	117z	72

Figure 2-6: A summary of frequency and normalized displacement results for materials of different density.

Material mass density is modified by using a custom material.

When you review the results in Figure 2-6, remember that while the material density changes, the model stiffness remains the same. Even though this exercise has no relevance to real life, it still provides interesting results. As we know, a structure vibrating in a given mode of vibration may be treated as a **Single Degree of Freedom** oscillator (SDOF) with a natural frequency expressed by: $\omega = \sqrt{\dfrac{k}{m}}$

In our case stiffness remains the same; therefore, the natural frequency is inversely proportional to the square root of mass. This is why a 100 fold increase in the mass causes the frequency to drop to one tenth of the original frequency.

Modal analysis normalizes the displacement results in such a way that the ratio of maximum displacements equals the inverse of the square root of the ratio of mass; compare displacement results in rows 1 and 3 in Figure 2-6.

Displacement results of modal analysis are so often misinterpreted that it may be better to show the results of modal analysis without colors, making it a deformation plot.

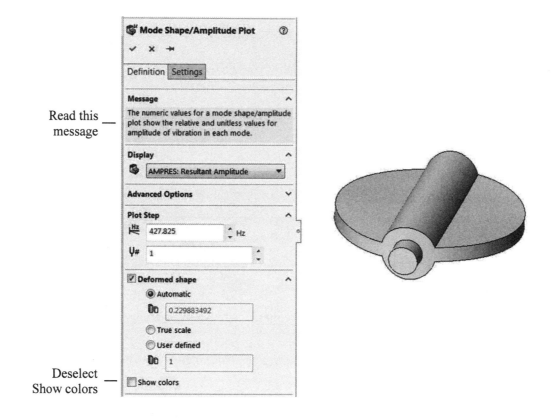

Read this message

Deselect Show colors

Figure 2-7: Plot of deformations in mode 1.

Deselecting colors eliminates color fringes and hides the color legend. This is often done to avoid misinterpretation of results.

The plot shows the shape of the first mode, material mass density 7800kg/m³.

Another way to avoid misinterpretation of results modal analysis is to normalize displacement results to 1 as shown in Figure 2-8.

Select
Normalize Mode Shape

Figure 2-8: Normalized plot of deformation in mode 1.

Normalizing displacement results to 1 is another method of avoiding misinterpretation of displacement results of modal analysis.

A stress analysis of the NOTCHED PLATE model requires careful preparation of the mesh. Create a **Static** study titled *01 stress* and define a mesh control on the notch face to assure the correct element size and turn angle (Figure 3-2).

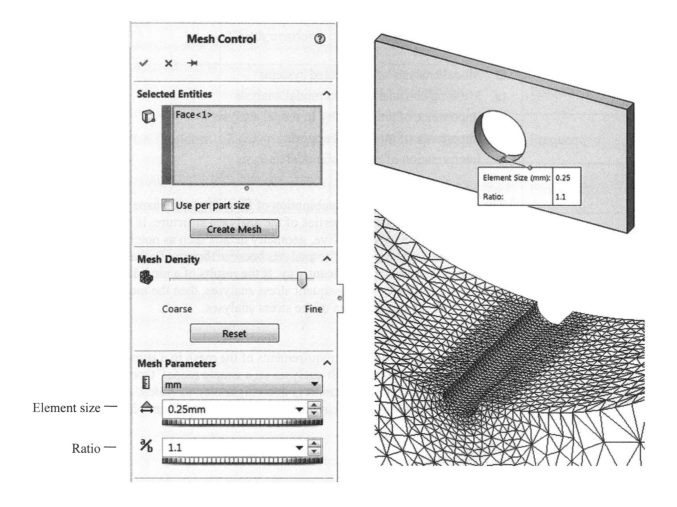

Figure 3-2: Mesh control defined on the face of the notch.

Element size on the controlled entity is 0.25mm; this is controlled independently of the global element size. Ratio defines the relative size of elements in consecutive layers in the transition zone between element size 0.25mm and global element size 5.27mm.

Define loads and restraints as shown in Figure 3-1 and obtain the solution. Next, copy study *01 stress* into a new study titled *02 stress*. Delete the mesh controls and obtain the solution with the default mesh. Von Mises stress results of both studies are shown in Figure 3-3.

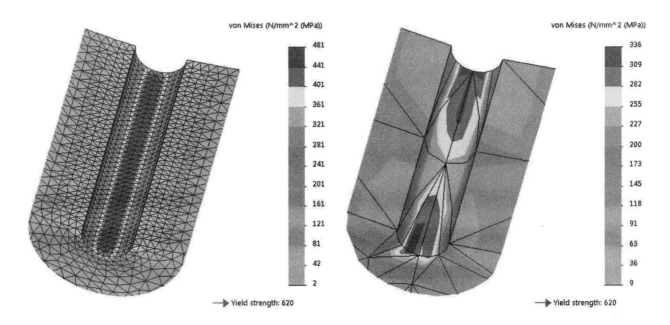

Study *01 stress*
Correct stress results produced
by a mesh with controls.
The maximum von Mises stress: 481MPa

Study *02 stress*
Incorrect stress results produced
by a default mesh.
The maximum von Mises stress: 336MPa

Figure 3-3: Von Mises stress results produced with the correct mesh (left) and incorrect mesh (right).

Using the default element size and no mesh controls produces highly distorted elements that give incorrect stress results. Plots use Section Clipping with a cylindrical surface aligned with the axis of the notch.

Compare the stress plots in Figure 3-3 and notice that the incorrect mesh returns an incorrect stress distribution pattern as well as von Mises stresses that are 25% lower as compared to the correct mesh. This quick review of stress analysis shows the importance of meshing small geometry details when stress results are sought.

We will now study the effects of the mesh on the results of modal analyses. Create two **Frequency** studies titled *03 modal* and *04 modal*. Define the same restraint as shown in Figure 3-1; don't define any load. In study *03 modal* use a mesh with the same controls as in study *01 stress*. In study *04 modal* don't use any mesh controls.

A summary of results of studies *03 modal* and *04 modal* is shown in Figure 3-4. The same figure also shows modal frequencies of a model without a notch. These results may be obtained by analyzing the model in configuration *01 no notch* using a default mesh; do this in study *05 modal*.

Mode number	Frequency [Hz]		
	Study *03 modal* Model with notch Mesh controls	Study *04 modal* Model with notch No mesh controls	Study *05 modal no notch* Model without notch No mesh controls
1	210.2	210.2	210.3
2	845.1	845.5	845.6
3	1274.4	1274.7	1275.3
4	1730.1	1730.1	1730.2
5	2825.9	2825.9	2826.1

Figure 3-4: Comparison of modal frequencies produced by the model with a notch and mesh controls, with a notch and no mesh controls, and without a notch.

Incorrect mesh or removal of the notch has next to no effect on the natural frequencies.

The results summarized in Figure 3-4 show that incorrect meshing of small details or removing those details has little effect on the natural frequency. Animate any mode to see that the modal shapes also remain the same.

Stresses may be strongly dependent on small **local** features which must be correctly represented in a mesh intended for stress analysis. Modes of vibration depend on structural stiffness which is a **global** model property. Incorrect meshing or removal of small features does not change stiffness in a significant way. This is the reason that modal analysis is insensitive to incorrect meshing of small features or to the removal of these small features all together.

The above statement applies only if a modal analysis is the only objective. If results of a modal analysis are used as pre-requisites to a subsequent vibration stress analysis, then all rigors of correct meshing apply to a modal analysis.

Stay with the *01 no notch* configuration open to perform a convergence analysis of natural frequencies. Copy study *05 modal no notch* into *06 modal no notch coarse* and mesh the model with a very coarse mesh using 25mm default element size. Then copy *05 modal no notch* into *07 modal no notch fine* and mesh the model with a very fine mesh using 2.5mm default element size. Frequency results in the first mode are shown in Figure 3-5.

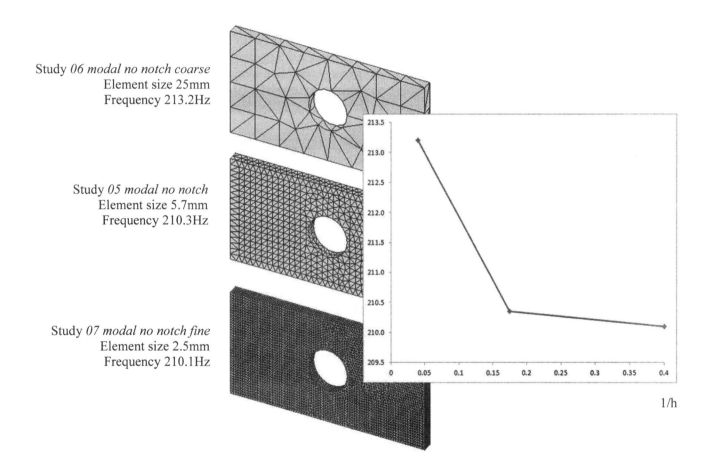

Study *06 modal no notch coarse*
Element size 25mm
Frequency 213.2Hz

Study *05 modal no notch*
Element size 5.7mm
Frequency 210.3Hz

Study *07 modal no notch fine*
Element size 2.5mm
Frequency 210.1Hz

1/h

Figure 3-5: Frequency of the first mode as a function of 1/h, where h is the element size.

Frequency drops with element size and converges to the asymptotic value corresponding to a continuous model (model before discretization).

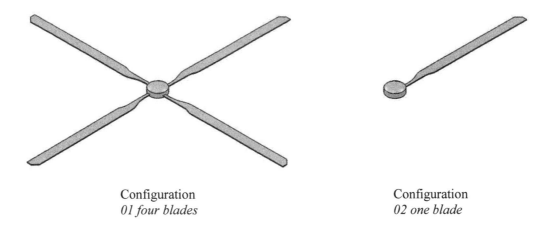

Configuration
01 four blades

Configuration
02 one blade

Figure 4-1: ROTOR model.

Modal analysis will be conducted on one blade.

Switch to *02 one blade* configuration and create a **Frequency** study *01 rotating*. Apply **Fixed** restraints as shown in Figure 4-2.

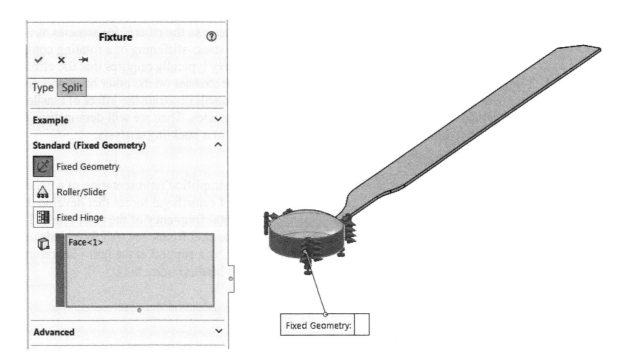

Figure 4-2: Restraints definition.

A fixed restraint is defined to the cylindrical surface. You may also apply it to the top and bottom face of the hub; it won't have any significant effect on the vibration of the blade.

Define a centrifugal load of 300RPM by selecting the cylindrical surface where the **Fixed** restraint has been defined. This cylindrical face uniquely defines the axis which is taken as the axis of rotation as shown in Figure 4-3.

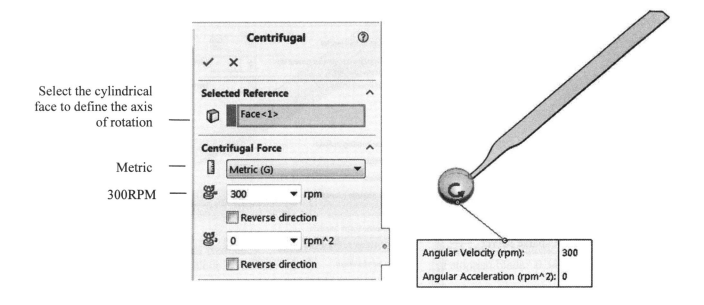

Figure 4-3: Centrifugal load definition.

A centrifugal load is defined by an angular rotation about the axis. Use metric units to define it in revolutions per minute (RPM). Do not define any angular acceleration.

Next, define a load as shown in Figure 4-8.

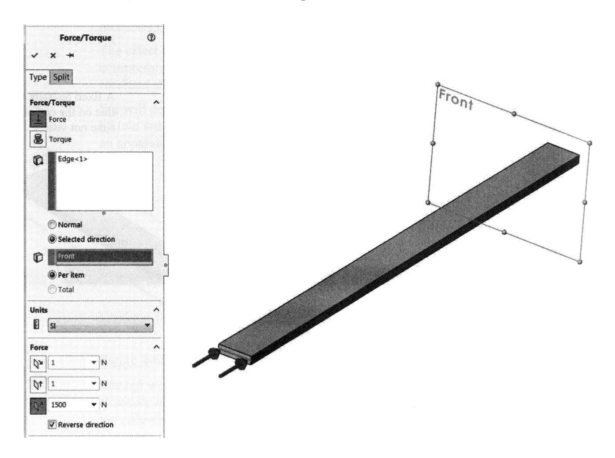

Figure 4-8: Load applied to the COLUMN model.

The load is applied to the same split line where the restraint on the side of the moving wedge has been applied.

Notice that on the loaded end, the restraint and load are both applied to the same entity (the split line), but in different directions. If the load was applied in the direction of this restraint it would be ineffective.

Mesh the model with the default element size and run the solution of the buckling study. The first buckling mode is shown in Figure 4-9 and the **Buckling Load Factor** (BLF) is 1.5755, meaning that according to the linear buckling model, buckling will happen at a load of 1575.5N.

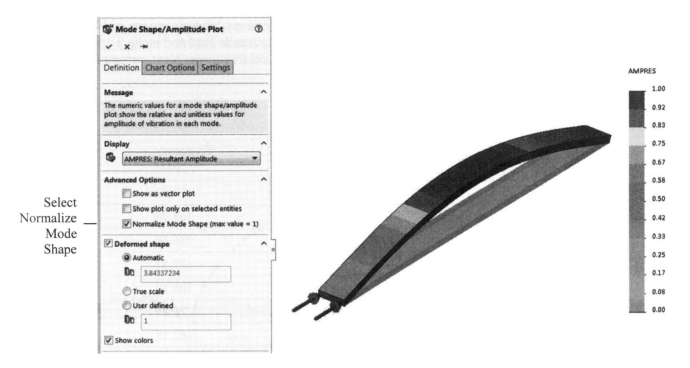

Select
Normalize
Mode
Shape

Figure 4-9: Buckled shape shown using displacement plot; undeformed model is superimposed on the deformed plot.

Even though numerical values are shown, their values may be used only to find a displacement ratio rather than absolute displacement. This is in close analogy to modal analysis. In this plot, the results have been normalized to 1.

The linear buckling analysis was necessary to establish the range of loads in the modal analysis with pre-stress. We'll now conduct a numerical experiment subjecting the model to different loads ranging from tensile to compressive to study the effect of pre-load on the fundamental natural frequency. The experiment will be conducted in twelve **Simulation** studies while the load is changed from a 1500N tensile load, to a compressive load causing buckling. A tensile load is denoted as positive, compressive as negative.

We should point out that numerical results presented in this experiment may differ slightly depending on the service pack used. In all cases the **Direct sparse solver** is used.

Summary of studies completed

Model	Configuration	Study Name	Study Type
ROTOR.sldprt	*02 one blade*	*01 rotating*	Frequency
		02 stopped	Frequency
COLUMN.sldprt	*Default*	*00 buckling*	Buckling
		01	Frequency
		02	Frequency
		03	Frequency
		04	Frequency
		05	Frequency
		06	Frequency
		07	Frequency
		08	Frequency
		09	Frequency
		10	Frequency
		11	Frequency
		12	Frequency
		12 fine	Frequency

Figure 4-14: Names and types of studies completed in this chapter.

5: Modal analysis - properties of lower and higher modes

Topics covered

- ❑ Modal analysis using shell elements
- ❑ Properties of lower and higher modes
- ❑ Convergence of frequencies with mesh refinement

Open the U BRACKET part model and notice that the bracket is modeled as a surface (Figure 5-1). The bracket is supported by two hinges along both ends.

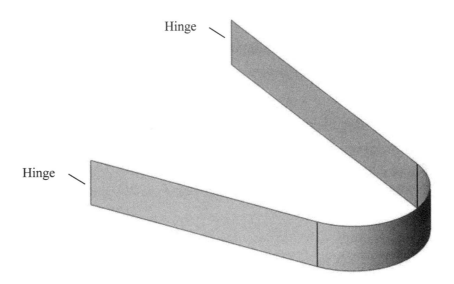

Figure 5-1: U BRACKET model supported by two hinges.

The geometry is represented by a surface. Therefore the thickness is missing from geometry.

Create a **Frequency** study *01 default mesh* and define a shell thickness of 2mm (Figure 5-2).

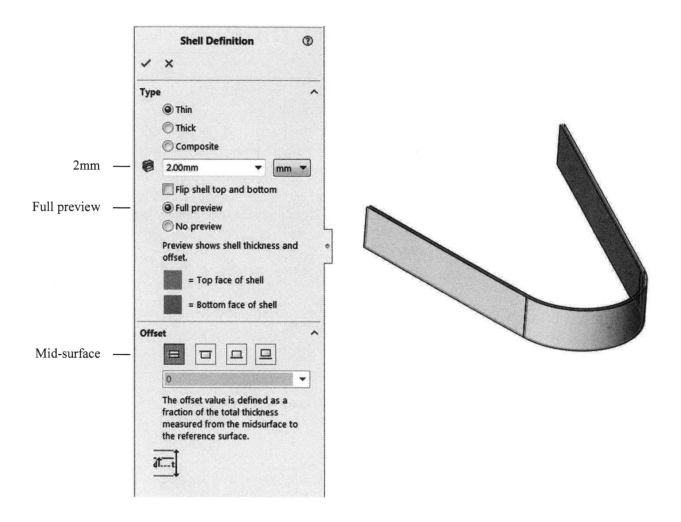

Figure 5-2: Shell thickness definition.

Mid-surface offset means that thickness is symmetric about the surface.

Define restraints as shown in Figure 5-3.

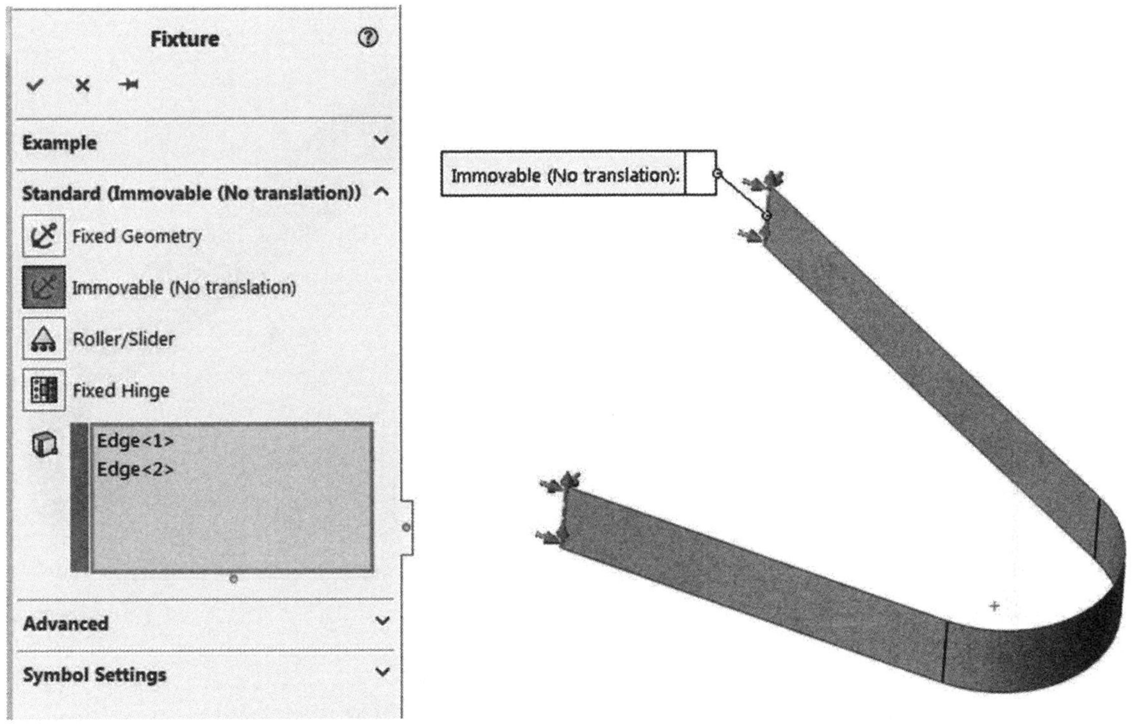

Figure 5-3: Hinge restraints.

Use Immovable restraints to allow hinging about the two short ends.

Notice that the **Fixture** window differentiates between **Fixed Geometry** and **Immovable (No translation)** restraints. This is because the model geometry is a surface, and **Simulation** recognizes it as a candidate for meshing with shell elements.

Mesh the model using the default settings (Figure 5-4).

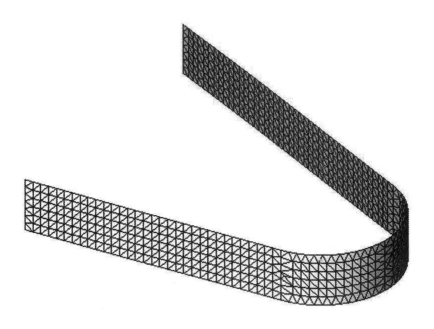

Figure 5-4: U BRACKET meshed with shell elements.

Colors (not visible in this black and white illustration) differentiate between top and bottom of shell elements. Here, the outside is meshed with tops; the shell's normal vector points from inside to outside the model. Refer to Figure 5-2 for the location of top and bottom of shells.

Shell element orientation is very important in stress analysis. Different stress results are reported on different sides of shell elements. In modal analysis, which does not calculate stress, shell orientation has no impact on results. This is of course correct only if modal analysis is our only objective. If it is followed by vibration analysis where stresses need to be analyzed, then the orientation of shell elements must be carefully controlled.

Obtain a solution for 10 modes of vibration and review the results as shown in Figure 5-5.

User defined scale
Colors deselected

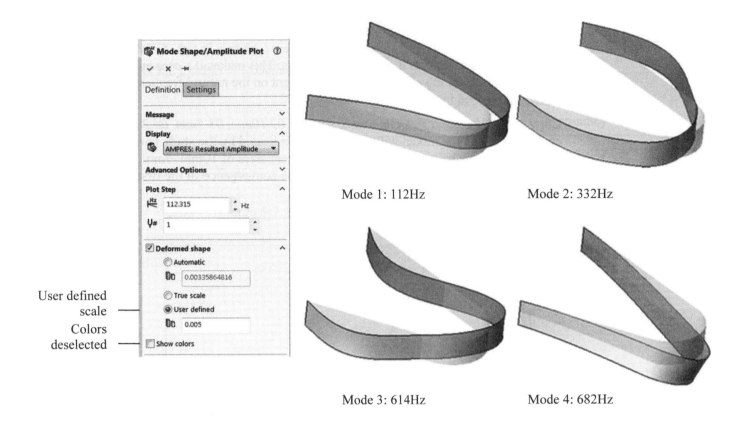

Mode 1: 112Hz   Mode 2: 332Hz

Mode 3: 614Hz   Mode 4: 682Hz

Figure 5-5: The first four modes of vibration of U BRACKET.

The undeformed model shape is shown along with the deformed model. To differentiate the undeformed shape from the modal shape, the scale of deformation has been adjusted individually for each mode. The Mode Shape window shows the deformation scale adjusted for Mode 1. Colors are deselected.

A review of shapes of all calculated modes reveals the following qualitative properties of lower and higher modes. Lower modes "find" the easiest way to move structure. Notice that the bracket vibrating in mode 1 hinges about supports; little deformation is present. Higher modes have progressively complex shapes while motion tends to be "more uniformly" distributed over the model.

The above findings indicate the following:

- Lower modes tend to maximize kinetic energy and minimize strain energy of a vibrating structure.
- Higher modes tend to maximize strain energy and minimize kinetic energy.

6: Modal analysis – mass participation, properties of modes

Topics covered

- Modal mass
- Modal mass participation
- Modes of vibration of axisymmetric structures
- Modeling bearing restraints
- Using modal analysis to find "weak spots"

Procedure

We'll analyze the SHAFT assembly model which is a simplified representation of a shaft with two gears, supported by two spherical bearings that allow for some angular displacement. Many small features that are not essential for modal analysis (chamfers, rounds, undercuts) are not present in this model. Those simplified or eliminated features would have to be modeled if we had to conduct stress analysis.

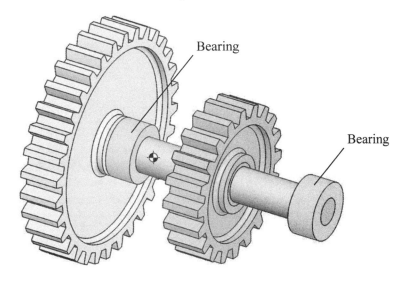

Figure 6-1: SHAFT supported by two spherical bearings.

Many features important for manufacturing, assembly and function are not included in this model. The model is shown in the 01 bearings configuration. The center of mass of the assembly is shown.

The objective of this exercise is to find the first few natural frequencies of the shaft supported by spherical bearings, and to compare these results with results of the shaft when unsupported. This exercise introduces vibration properties of axi-symmetric structures, modal mass participation and the effect of restraints on the mass participating in vibration. The major modeling consideration is modeling supports that allow for angular displacement present in spherical bearing supports.

The SHAFT assembly model has three configurations: *01 bearing, 02 no bearings,* and *03 spherical bearing faces*. Configuration *01 bearings* serves only to illustrate the problem. Notice that applying restraints to cylindrical faces of two bearings would eliminate rotations allowed by the spherical bearings. The difference between a fixed bearing support and a spherical bearing support is easy to explain if we consider modes of vibration of a beam with fixed supports and simple supports.

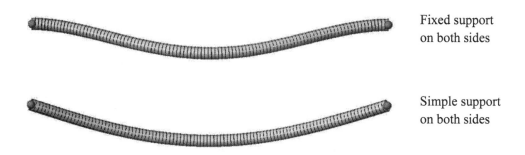

Fixed support on both sides

Simple support on both sides

Figure 6-2: Bending mode shape of a beam with fixed supports (top) and simple supports (bottom).

The ends of the beam with simple supports are able to rotate.

The model used to create plots in Figure 6-2 is called BEAM DEMO. It is not directly related to SHAFT exercise.

The difference between fixed and simple supports is easy to model using beam elements, as shown in the BEAM DEMO model. These two types of supports are differentiated in BEAM DEMO by selecting either **Fixed** or **Immovable** restraints. Review this model and notice that the simple support on both sides requires an additional restraint to eliminate a rigid body motion; rotation about beam axis must be eliminated.

In the case of the SHAFT model where restraints can't be applied to one point, we'll need to use a **Bearing Support** as shown in Figure 6-3.

A short note on the terminology issue: the pop-up menu in Figure 6-4 shows command **List Resonant Frequencies**. This should be called List Modal Frequencies because a resonant frequency and modal frequency are not the same even though these terms are often used interchangeably. The modal frequency is found in the absence of damping and excitation; it is a property of structure. The resonant frequency is affected by damping and requires an excitation to manifest itself.

Frequency and shape are not the only factors defining mode of vibration. Another very important characteristic is the **Mass Participation**; it indicates what percentage of the total mass participates in vibration in the given mode. Follow steps in Figure 6-5 to look into **Mass Participation** in all 16 modes found in the SHAFT assembly.

Change the model configuration to *02 no bearings*, create a **Frequency** study titled *01 bearing support* and define **Bearing Fixture** restraints as shown in Figure 6-3.

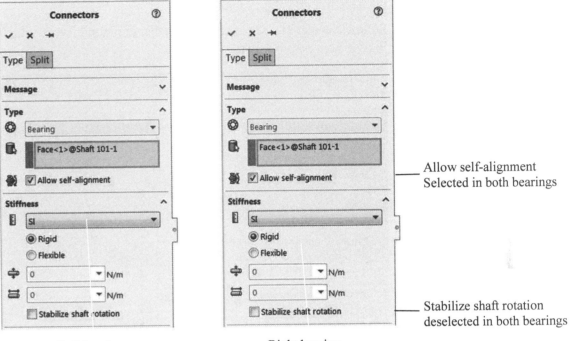

Left bearing   Right bearing

Allow self-alignment
Selected in both bearings

Stabilize shaft rotation
deselected in both bearings

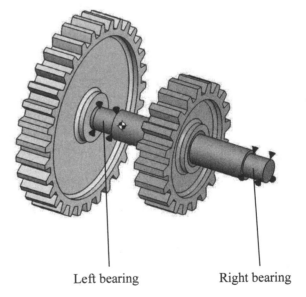

Left bearing   Right bearing

Figure 6-3: Definition of Bearing Support.

Bearing supports must be defined individually for each side. Definitions of left bearing support and right bearing support are identical.

Since the shaft rotation is not restrained, the model will have one **Rigid Body Mode** – rotation about its axis.

Specify 16 modes in the study properties. Mesh the model with default element size and obtain solution.

Follow steps in Figure 6-4 to display a list of natural frequencies.

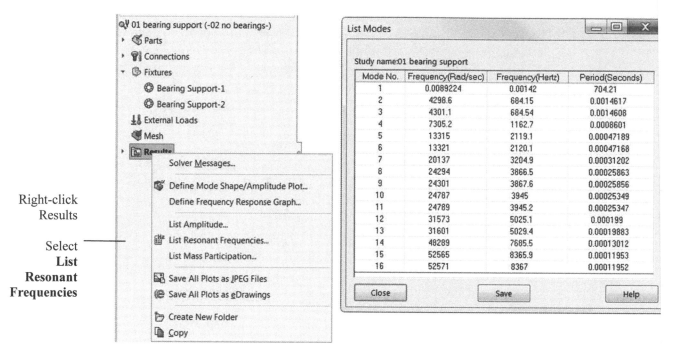

Figure 6-4: List Modes table shows the natural frequencies.

Frequencies are listed in the ascending order.

Modal frequencies in Figure 6-4 are shown using three units: rad/s, Hz, and s.

The relation between frequency ω *(rad/s)* and f *(Hz)* is: $\omega = 2\Pi f$

The relation between frequency f *(Hz)* and vibration period T is: $T = 1/f$

Review all three graphs that can be constructed with **Define Frequency Response Graph**. Three graphs may be created:

- Frequency vs. Mode number

- Frequency vs. effective mass participation factor (EMPF)

- Frequency vs. cumulative effective mass participation factor (CEMPF)

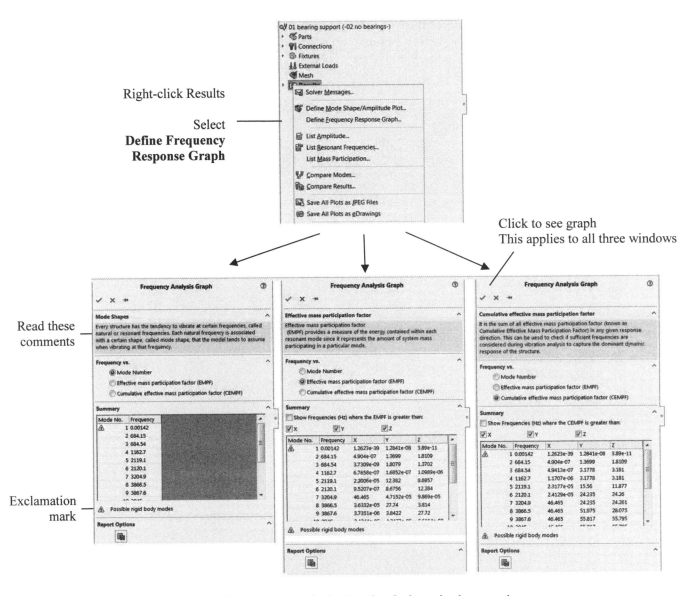

Figure 6-5: Frequency Analysis Graph windows in three options.

The table in Frequency vs. Mode number window is a repetition of Figure 6-4. Each window can be shown as a graph and/or saved in the csv format for processing in Excel. Notice an exclamation mark by mode 1 indicating this is a Rigid Body Mode (RBM).

Now, examine how many modes need to be considered to reach the **Cumulative effective mass participation factor** (CEMPF) over 80% of the total mass; follow steps shown in Figure 6-6.

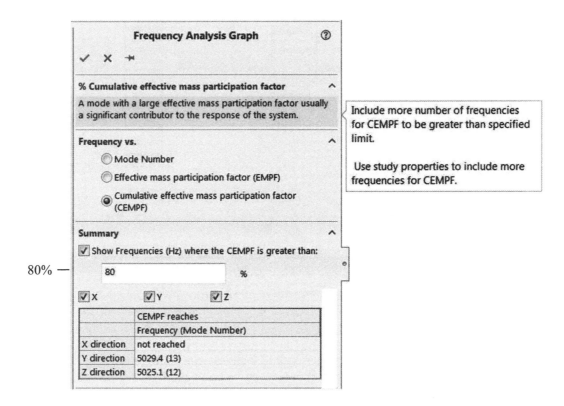

Figure 6-6: 13 modes are required to reach the CEMPF factor in X direction ; 12 modes are required to reach the CEMPF in Z direction. More than 16 modes would be required to reach the CEMPF over 80% in Z direction.

Bending modes in Y and Z directions are coupled modes.

Animate the modes with significant mass participation in X direction. These are modes 7 and 14; the numbering includes the Rigid Body Mode which is counted as mode 1.

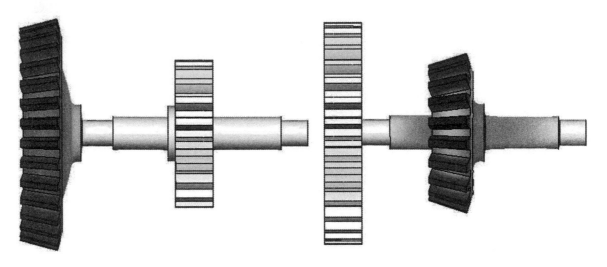

Mode 7: 3205Hz   Mode 14: 7686Hz

Figure 6-7: Shapes and frequencies of mode 7 and mode 14. These modes have significant mass participations in X direction.

Mode 7 is associated with deformation of the large gear; mode 14 is associated with deformation of the small gear.

Animation of modes 7 and 14 clearly shows mass moving in X direction. Notice that the shaft is unlikely to be excited in X direction. Therefore, these modes may never manifest themselves, especially mode 14 which has a very high frequency.

Shapes and frequencies of mode 8 and mode 9 are shown in Figure 6-8.

Mode 8: 3866.5.6Hz   Mode 9: 3867.6.5Hz

Figure 6-8: Shapes and frequencies of mode 8 and mode 9; both are bending modes.

Mode 8 is associated with deformation of shaft in XZ plane; it has significant mass participation in Z direction.

Mode 9 is associated with deformation of shaft in XY plane; it has significant mass participation in Y direction.

Animate modes 2 and 3, 5 and 6, 12 and 13 before proceeding.

We may use the modal results shown in Figure 6-8 to make important observations about modal frequencies and shapes of the SHAFT model.

Look at four pairs of modes: 2 and 3, 5 and 6, 8 and 9, 12 and 13. All these modes are associated with rotation about the centers of the bearings; the results are summarized in Figure 6-9.

	Mode number	Frequency [Hz]
Pair 1	2	684.15
	3	684.54
Pair 2	5	2119.1
	6	2120.1
Pair 3	8	3866.5
	9	3867.6
Pair 4	12	5025.1
	13	5029.4

Figure 6-9: Frequencies of four pairs of modes identified as bending modes in mutually orthogonal planes.

Each pair has almost the same frequency. Animation indicates that modal shapes are identical but rotated by 90° about the X axis.

Identical modal shapes and almost identical frequencies are not coincidental. Repetitive modes with the plane of vibration rotated by 90° characterize modal results of axi-symmetric structures. These modes are called coupled modes. A small difference in the numerical value of frequency is caused by the presence of teeth and by the discretization error. Teeth make the model not perfectly axi-symmetric; discretization error also makes the stiffness of the finite element model not perfectly axi-symmetric.

Animate the modal shapes of coupled modes to confirm that the shapes are indeed identical but rotated by 90° (Figure 6-8).

Next, animate mode 4 which has no significant mass participations in any direction. Shape and frequency of mode 4 are shown in Figure 6-10.

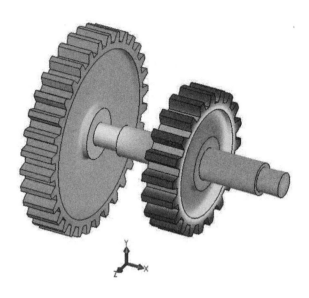

Mode 4: 1162.7Hz

Figure 6-10: Shape and frequency of mode 4; this mode has no significant mass participation in any of the principal directions.

This is a torsional mode; two gears rotate in the opposite directions about the axis of the shaft while the connecting shaft experiences torsion.

Finally, animate mode 1 which is the **Rigid Body Mode**; its modal shape shows that SHAFT assembly rotates as a rigid body about its axis (Figure 6-11).

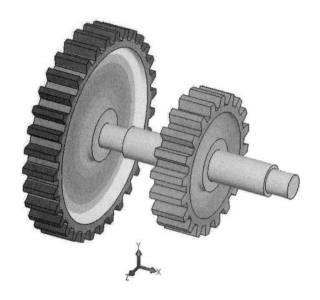

Mode 1: 0.00142Hz

Figure 6-11: Shape and frequency of the Rigid Body Mode 1.

Frequency 0Hz is assigned to rigid body modes by solver. The result doesn't show exactly 0Hz because of discretization error.

Shapes of the torsional mode 4 (Figure 6-10) and the rigid body mode 1 (Figure 6-11) look the same in black and white illustrations; you need to animate them in **Simulation** to see differences. When animated, the shapes of modes 1 and 4 are easy to observe because of the teeth. This is why the assembly model SHAFT features schematically modeled teeth. Modal shapes 1 and 4 would be impossible to observe if teeth were not modeled.

In continuation of the SHAFT exercise we will compare the shape of frequency of the first bending mode between SHAFT assembly supported by bearings and unsupported SHAFT assembly. Copy study *01 bearing support* into *02 no supports*, delete or suppress bearing support and obtain solution; the solution will show six rigid body modes. The first elastic model will be mode 7 which is a torsional mode; mode 8 will be the first bending mode. Therefore, the comparison needs to be done between mode 2 of the model with bearing supports and mode 8 of the unsupported model.

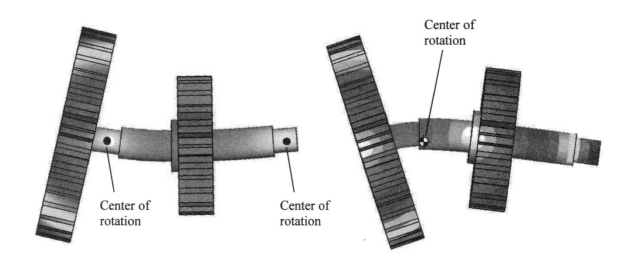

Spherical bearings support

Mode 2
Frequency: 684Hz

Vibration takes place about the spherical bearings

No supports

Mode 8
Frequency: 876Hz

Vibration takes place about the center of mass

Figure 6-12: Comparison of the first bending mode of SHAFT with bearing supports and unsupported.

The center of mass symbol may be inserted into assembly as reference geometry.

In the absence of supports, vibration always takes place about the center of mass. Removal of the bearing supports affects the model stiffness, and mass participation. The combined effect of both changes results in the increase of the frequency of the first bending mode.

The SHAFT model can be also used to illustrate that modal analysis provides a qualitative review of "weak spots." While absolute displacements can't be found in modal analysis, the relative displacements are valid. Return to configuration *01 bearing support* and animate bending mode 2 and bending mode 8 of model supported by spherical bearings. Observe bending of the small cantilever shaft holding the large gear. This animation reveals a potential design fault: the heavy gear held by a soft cantilever shaft (Figure 6-13).

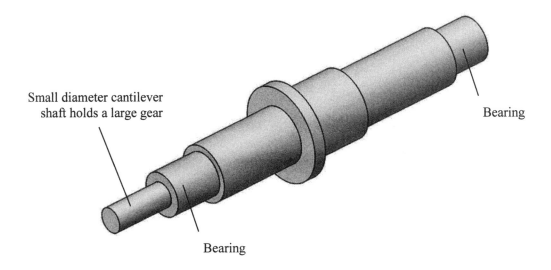

Small diameter cantilever shaft holds a large gear

Bearing

Bearing

Figure 6-13: Potential design flaw identified by modal analysis: a soft cantilever beam supports a heavy gear.

The split face used to define a bearing support near the large gear is not shown in this illustration.

The flexibility of the large gear sitting on the small cantilever shaft may cause alignment problems, noise and premature wear. Therefore, based on the results of this analysis, the designer may decide to move the bearing to the end of the shaft.

We used a **Bearing Support** to provide the model with the ability to perform rotations about the bearings even though bearings themselves were not included in the model. The same support may be represented if bearings are added to the assembly and their outside face is modeled spherically as in assembly configuration *03 spherical bearing faces.*

Figure 6-14 shows the **On Spherical Faces** restraints defined for both bearings; this is equivalent to the **Bearing Support** with **Allow Self-alignment** selected as shown in Figure 6-3.

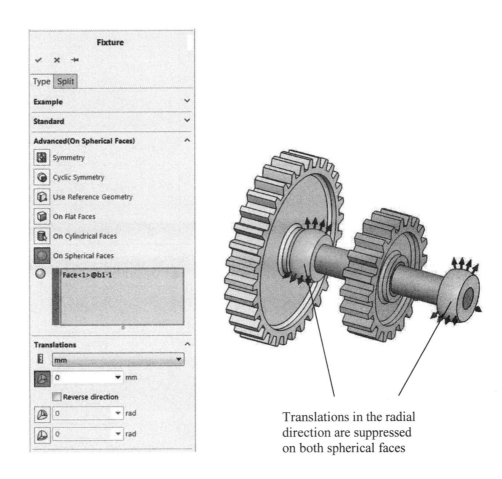

Translations in the radial direction are suppressed on both spherical faces

Figure 6-14: Bearing supports modeled using restraints defined in local spherical coordinate systems.

This restraint has to be defined individually for each bearing.

Switch to *03 spherical bearing faces* configuration and create **Frequency** study *03 spherical*. Analyze the model with restraints defined as shown in Figure 6-14 and verify that the results are close to those obtained using **Bearing Supports**. The slight difference is due to the fact that the model shown in Figure 6-14 takes into consideration the inertial effects of the bearings.

Summary of studies completed

Model	Configuration	Study Name	Study Type
BEAM DEMO.sldprt	Default	*01 Fixed*	Frequency
		02 Immovable	Frequency
SHAFT.sldasm	*02 no bearings*	*01 bearing support*	Frequency
		02 no support	Frequency
	03 spherical bearing faces	*03 spherical*	Frequency

Figure 6-15: Names and types of studies completed in this chapter.

7: Modal analysis – mode separation

Topics covered

- ❑ Modal analysis with shell elements
- ❑ Modes of vibration of symmetric structures
- ❑ Symmetry boundary conditions in modal analysis
- ❑ Anti-symmetry boundary condition in modal analysis

Procedure

The CAR model is a simplified representation of a car body; we'll use it to find the first torsional mode of vibration. Frequency of the first torsional mode is an important factor in vehicle handling and ride quality. Modern uni-body sedans have the first torsional frequency in the range 25-30Hz. Convertibles have "softer" bodies and, consequently, a lower torsional frequency of 10-15Hz.

Due to a schematic representation of the car body design, the CAR model will not return realistic numerical values of natural frequencies. Still, it will allow us to review important modeling techniques used in modal analysis.

Review the CAR model in its two configurations *01 full* and *02 half* (Figure 7-1).

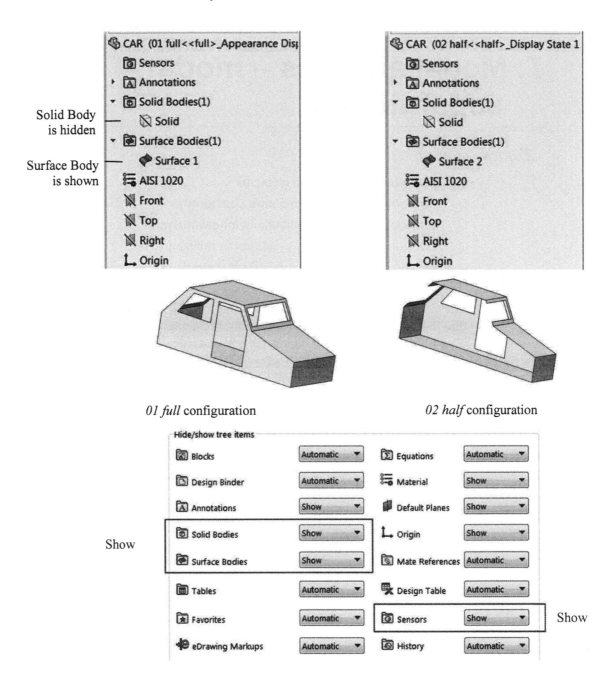

01 full configuration   02 half configuration

Figure 7-1: Controlling display of Solid Bodies and Surface Bodies. The CAR model consists of Solid Body and Surface body; the Solid Body is hidden.

To show the Solid Bodies and Surface Bodies folders in the SOLIDWORKS Feature Manager Design Tree, right click anywhere in the Feature Manage window and use the pop-up window to show these items. Do not confuse show/hide of Solid Bodies folder and Surface Bodies folder in the Feature Manager Design Tree with show/hide of actual Solid or Surface bodies which are held in these folders. While modifying visibility of tree items, make Sensors show too; this will be used later in this book.

Activate the *01 full* configuration and create a **Frequency** study called *01 full*. In the study properties, request 12 modes to be found. Exclude the **Solid Body** from the analysis as shown in Figure 7-2.

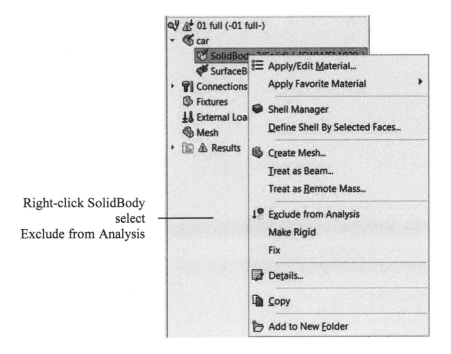

Figure 7-2: Solid Body must be excluded from the analysis to avoid meshing of the Solid Body.

Even though the Solid Body has been hidden in the CAD model, it still has to be removed from analysis.

Right-click **Surface Body**, select **Edit Definition** and enter a shell element thickness of 50mm (we do not attempt to model a car body realistically).

Obtain the solution using the default element size and review frequencies of all modes; follow steps shown in Figure 7-3.

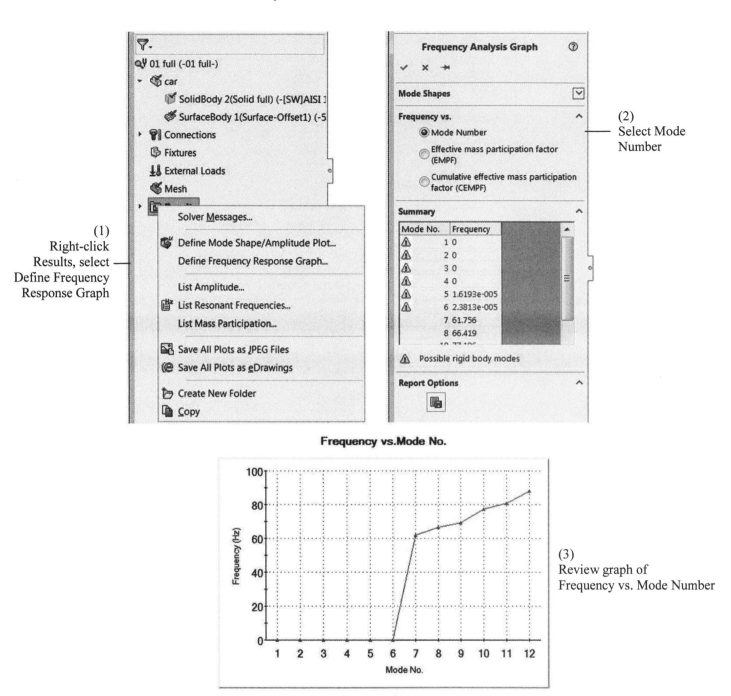

Frequency vs.Mode No.

Figure 7-3: In the absence of restraints, the first six modes are Rigid Body Modes with frequencies very close to or equal to 0Hz.

Modes 5 and 6 are not exactly zero due to discretization error. Frequency vs. Mode Number graph has been created automatically. Lines between points have no meaning; therefore, points in the graph should not be connected. Export this graph to Excel and re-construct it as a bar graph.

Animate the six elastic modes (modes 7-12) and notice that the second elastic mode of vibration, which is mode number 8, is the torsional mode we are looking for. Also, observe that the deformed shapes are either symmetric or anti-symmetric (Figure 7-4).

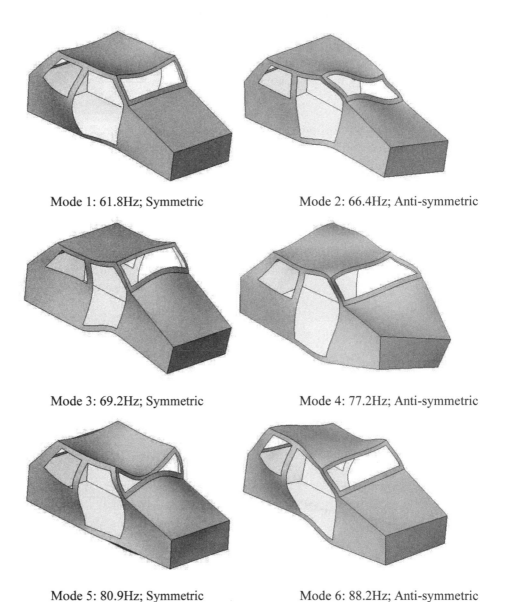

Mode 1: 61.8Hz; Symmetric   Mode 2: 66.4Hz; Anti-symmetric

Mode 3: 69.2Hz; Symmetric   Mode 4: 77.2Hz; Anti-symmetric

Mode 5: 80.9Hz; Symmetric   Mode 6: 88.2Hz; Anti-symmetric

Figure 7-4: Modal shapes and frequencies of the first six elastic modes.

It is a coincidence that the symmetric and anti-symmetric modes are alternating.

The property of a modal shape to be either symmetric or anti-symmetric applies to all structures with a plane of symmetry. You may want to return to the SHAFT model in chapter 6 to confirm this.

Having found this important property of vibration of symmetric structures, we may ask if analysis can be performed on one half of the model. If so, and what boundary conditions should be defined? We will investigate this using the CAR model in the *02 half* configuration. We will first run a modal analysis with symmetry boundary conditions, then a modal analysis with anti-symmetry boundary conditions.

Switch to configuration *02 half*, create a **Frequency** study titled *02 sym*, exclude **Solid Body** from the analysis and define the same shell element thickness as before (50mm). Define **Symmetry Boundary Conditions** as shown in Figure 7-5.

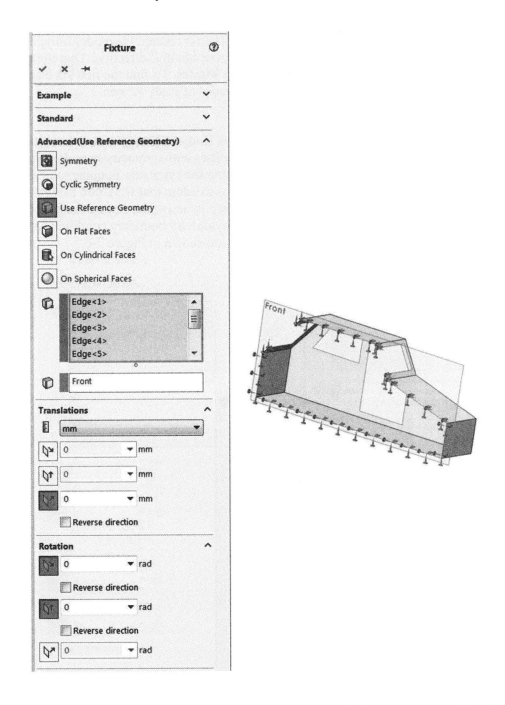

Figure 7-5: Symmetry boundary conditions applied to the edges of the shell element model.

Symmetry boundary conditions leave the model with 3 Rigid Body Motions.

Symmetry boundary conditions must be defined in terms of six degrees of freedom: three translations and three rotations. This is because the model uses shell elements which have six degrees of freedom per node.

The anti-symmetry boundary conditions do not fully restrain the model either. A summary of symmetry and anti-symmetry boundary conditions is shown in Figure 7-7.

	Symmetry BC	Anti-symmetry BC
X translation	Free	Fixed
Y translation	Free	Fixed
Z translation	Fixed	Free
X rotation	Fixed	Free
Y rotation	Fixed	Free
Z rotation	Free	Fixed

Figure 7-7: Summary of Rigid Body Motions in a model with symmetry and anti-symmetry boundary conditions.

Notice that what is fixed in symmetry boundary conditions is free in anti-symmetry boundary conditions and vice versa.

Obtain the solution of *03 anti-sym* and proceed to review of results of both studies. Animate the first three modes in study *02 sym* and in study *03 anti-sym* and confirm that they are **Rigid Body Modes** corresponding to rigid body displacements identified in Figure 7-7.

Animate the elastic modes found in study *02 sym*; these are Modes 4-9 and notice that all shapes are symmetric. Next, animate Modes 4-9 found in study *03 anti-sym* and notice that all shapes are anti-symmetric.

Symmetry boundary conditions eliminate anti-symmetric modes and anti-symmetry boundary conditions eliminate symmetric modes. Performing modal analysis on one half of the model first with symmetry boundary conditions then with anti-symmetry boundary conditions is called the **Modal Separation** technique.

One half of a symmetric model can be used to extract all modes of vibration by combining results of a modal analysis with symmetry boundary conditions with the results of modal analysis of the same model with anti-symmetry boundary conditions as it is schematically shown in Figure 7-8.

Study name	01 full		02 sym All modes are symmetric		03 anti-sym All modes are anti-symmetric	
	Mode No.	Hz	Mode No.	Hz	Mode No.	Hz
	1	0	1	0	1	0
	2	0	2	0	2	0
	3	0	3	0	3	0
	4	0	4	62	4	65
	5	0	5	68	5	76
	6	0	6	81	6	87
sym	7	62	7	88	7	91
anti-sym	8	66	8	96	8	145
sym	9	69	9	117	9	164
anti-sym	10	77				
sym	11	81				
anti-sym	12	88				

Figure 7-8: Summary of results of studies *01 full*, *02 sym*, and *03 anti-sym*.

Borders around the first six modes of the unsupported full model (study 01 full) and around the first three modes of the partially supported models (studies 02 sym and 03 anti-sym) indicate Rigid Body Modes.

The solutions of study *01 full* may be obtained by merging solutions of study *02 sym* obtained with symmetry boundary conditions and *03 anti-sym* obtained with anti-symmetry boundary conditions.

The **Modal Separation** technique may be also used to directly find the first torsional mode, which is the first elastic mode in the model with anti-symmetry boundary conditions.

Summary of studies completed

Model	Configuration	Study Name	Study Type
CAR.sldprt	*01 full*	*01 full*	Frequency
	02 half	*02 sym*	Frequency
		03 anti-sym	Frequency

Figure 7-9: Names and types of studies completed in this chapter.

8: Modal analysis of axi-symmetric structures

Topics covered

❑ Modes of vibration of axi-symmetric structures
❑ Repetitive modes
❑ Solid and shell element modeling

In this chapter we analyze an axi-symmetric model. The model VASE comes in two configurations *01 solid* and *02 shell*. The wall thickness in *01 solid* is 2mm. The surface in *02 shell* has been offset 1mm into the material. Either configuration may be used for analysis of modes of vibration of this axi-symmetric model. We will use *02 shell* which is intended for meshing with shell elements. If you use *01 solid*, make sure to use a sufficiently small element size to avoid excessive element distortion. Your results will be slightly different because of the different modeling approach and different discretization error.

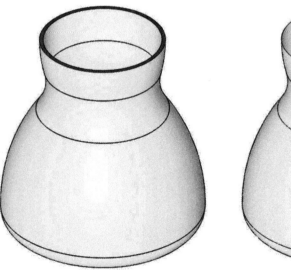
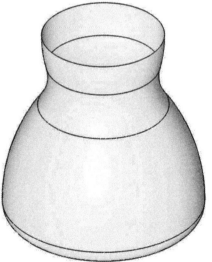

Configuration *01 solid*
Solid body can be meshed
with solid elements

Configuration *02 shell*
Surface body can be meshed
with shell elements

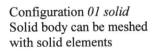

Figure 8-1: VASE model in two configurations.

Either configuration may be used for modal analysis.

Make sure the model is in the *02 shell* configuration and create a **Frequency** study titled *01 modal*. Verify that the material, **Glass**, has been transferred from the CAD model.

You don't have to exclude the **SolidBody** because it has been deleted from the model in *02 shell* configuration.

Define the shell thickness as 1mm (Figure 8-2).

(1)
Right-click
Surface folder —
Select Edit
Definition

(2)
Select
Thin definition;
Enter 1mm

(3)
Select
Flip shell top and
bottom
(for practice only)

(4)
Select
Middle Surface

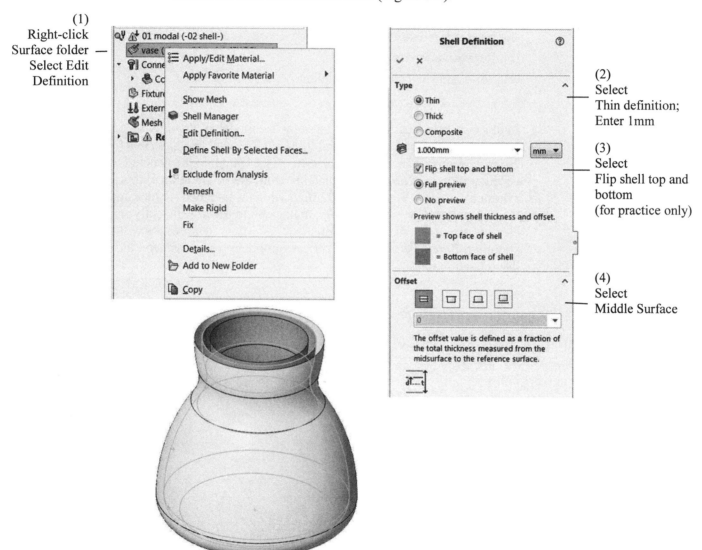

Figure 8-2: Thickness defined for Surface Body.

Middle Surface returns symmetric placement of thickness on both sides of the surface. The above image has been created using 5mm shell thickness; this is for better clarity of this illustration only.

Top and bottom of shells may be flipped as shown in Figure 8-2 where this is done only for practice. Shell element orientation has no importance in a model intended for modal analysis only.

Do not define any restraints to VASE model.

Mesh the VASE model using standard mesh 4mm element size and obtain solution for 12 modes. Vibration shapes and frequencies of the first two elastic modes are shown in Figure 8-3.

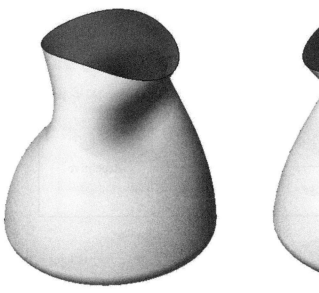 

Mode 1: 893.12Hz   Mode 2: 893.22Hz

Figure 8-3: The first two modes of vibration.

The frequency of these two modes is practically the same; the minute difference is caused by discretization error. The shape of both modes is identical but rotated about the axis of symmetry. The unsupported model has six Rigid Body Modes; therefore, the above mode 1 is numbered in results as mode 7 and mode 2 as mode 8.

Repetitive modes characterize results of modal analysis of axi-symmetric structures. More repetitive modes are listed in Figure 8-4.

Summary of studies completed

Model	Configuration	Study Name	Study Type
VASE.sldprt	02 shell	01 modal	Frequency
		02 modal	Frequency

Figure 8-6: Names and types of studies completed in this chapter.

9: Modal analysis – locating structurally weak spots

Topics covered

- ❑ Modal analysis with beam elements
- ❑ Modes of vibration of symmetric structures
- ❑ Using results of modal analysis to identify potential design problems
- ❑ Frequency shift

Procedure

Open the BAJA FRAME part (Figure 9-1). This model is an early iteration of a tubular frame designed by students of Western Engineering for the SAE BAJA competition. At an early stage of the design process, loads and restraints are largely unknown. Since modal analysis may be run without any restraints, an insight offered by modal analysis into structural properties of this design is particularly valuable.

Review the BAJA FRAME model and notice that all trims of Structural Members have been suppressed. Nodes are constructed in such a way that the center lines of the tubes all meet in one point. This was done to facilitate meshing with beam elements; manufacturing considerations will later require modifications to this CAD model.

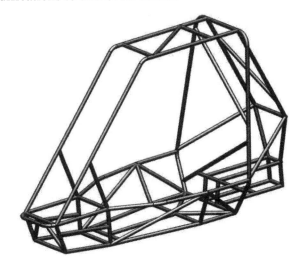

Figure 9-1: An early design iteration of Western Engineering's SAE BAJA frame.

The SOLIDWORKS model uses structural members suitable for meshing with beam elements.

Numerical values of modal frequencies are of little use except to find out which portions of the frame are "soft." Notice that the BAJA FRAME will be stiffened by installing the engine, transmission, suspension, steering and driver's seat. At the same time these components will also add mass. Therefore, at this point the modal analysis provides only qualitative results.

Based on these qualitative results, Western Engineering SAE BAJA team decided to reinforce locations indicated in Figure 9-2. The redesign was important for the driver's safety because these locations are critically important in the case of a roll-over.

We'll use the BAJA FRAME model to introduce **Frequency Shift**. This solver option allows for calculating frequencies clustered around the specified frequency value. It may be used to eliminate calculation of 0Hz frequencies in models where **Rigid Body Modes** are present.

Copy the completed study into two more studies: *Study 2* and *Study 3* and try out the **Frequency Shift** option using the settings shown in Figure 9-3 and Figure 9-4.

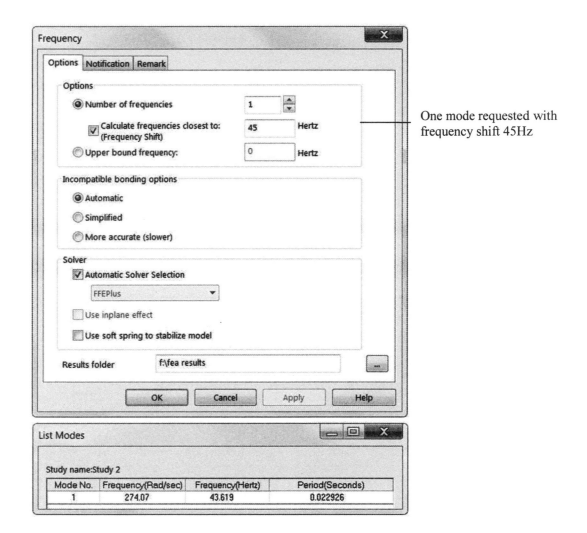

Figure 9-3: Properties and results of *Study 2* with one mode requested and frequency shift 45Hz. One mode with frequency closest to 45Hz has been found.

Direct sparse solver must be used when Frequency Shift is requested. Solver selection may be done automatically.

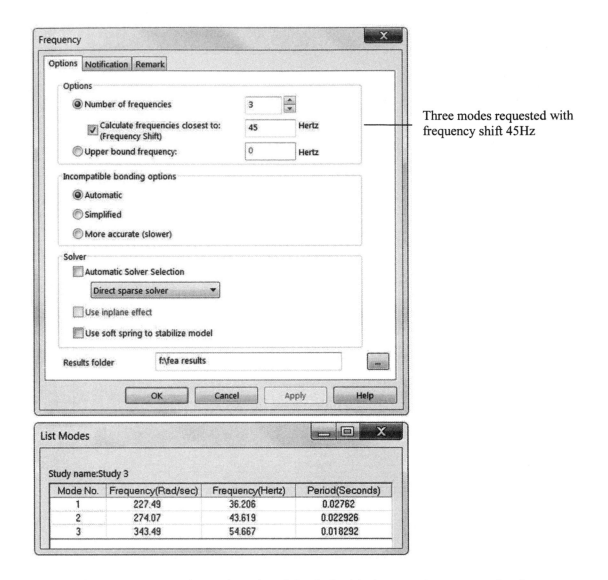

Three modes requested with frequency shift 45Hz

Figure 9-4: Properties and results of *Study 3* with three modes requested and frequency shift 45Hz. Three modes with frequencies closest to 45Hz have been found.

Modes with frequency below and above the Frequency Shift are found.

A **Frequency Shift** of 45Hz with just one mode requested finds the closest mode (43.2Hz) which is the torsional mode (Figure 9-3). A **Frequency Shift** with three modes requested finds modes 36.0Hz, 43.2Hz, and 54.3Hz (Figure 9-4). In both cases all rigid body modes are eliminated from the solution.

Summary of studies completed

Model	Configuration	Study Name	Study Type
BAJA FRAME.sldprt	*Default*	*Study 1*	Frequency
		Study 2	Frequency
		Study 3	Frequency

Figure 9-5: Names and types of studies completed in this chapter.

10: Modal analysis - a diagnostic tool

Topics covered

- ❑ Modal analysis used to detect problems with restraints
- ❑ Modal analysis used to detect connectivity problems
- ❑ Rigid Body Motion of assemblies

The ability of modal analysis to detect Rigid Body Motions can be used to diagnose modeling problems such as insufficient restraints or disconnected parts. If static analysis fails with the "insufficient restraints" or similar error message, you may still run a modal analysis on the same model and identify the direction(s) of movement corresponding to the 0Hz mode(s). This will identify the direction(s) where restraints are missing.

Procedure

The objective of the analysis of the PLIERS assembly is to find stresses in the arms of the pliers squeezing a plate as shown in Figure 10-1.

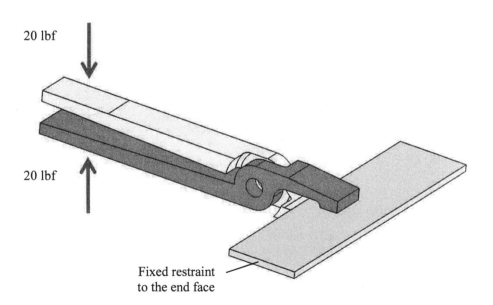

Figure 10-1: A plate being held by pliers is modeled in the PLIERS assembly.

The arms are loaded with a normal force of 20lbf applied to split faces as shown. Two arms of the pliers are connected with a pin connector; the pin is modeled as a pin connector; the pin connector is hidden in this illustration.

Open the assembly model PLIERS; make sure it is in *01 error* configuration. The assembly model has a static study *01static* and frequency studies *02 modal* and *03 modal* defined. The two arms are connected with a pin connector and loaded with 20lbf each; the plate has a fixed restraint defined to one side face (Figure 10-1).

An attempt to solve **Static** study *01 static* produces error messages shown in Figure 10-2.

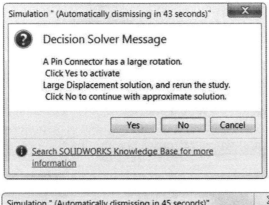

Select No to display the lower window

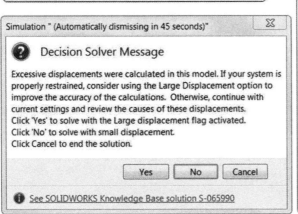

Select No to complete solution despite error

Messages displayed during solution

Solver Message window invoked by right-click Results folder

Figure 10-2: Error message issued by solver due to Rigid Body Motion detected in the model.

If you select No in the top window, the bottom window will be displayed. The same messages may be displayed later in Solver Message window.

The lack of "adequate fixtures" reported by both solvers is related to the presence of **Rigid Body Motions** (RBM) in the model; we will use modal analysis to find them. Knowing **the Rigid Body Motions** will help us identify problems with restraints.

Move to **Frequency** study *02 modal;* it has the same **Fixtures** and **Connectors** as the failed study *01 static*. There is no load in *02 modal* study.

At this point we know that there are unintentional RBMs in the model but we don't know how many are there; define the number of frequencies in study *02 modal* as 12. Obtain the solution and create **Frequency vs. Mode Number** graph to identify the RBMs in the model (Figure 10-3).

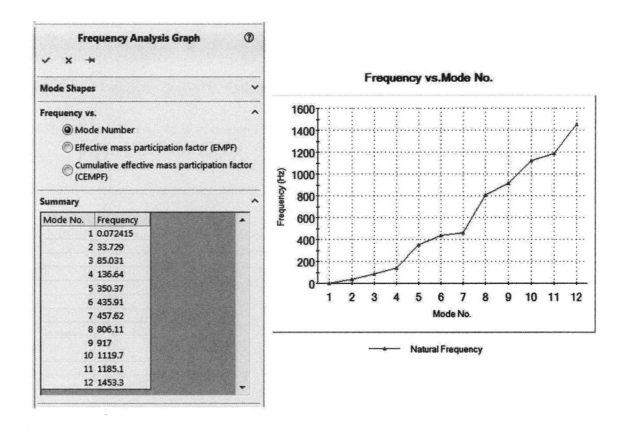

Figure 10-3: Modes 1 has near zero frequency.

There is one RBM in the model; the frequency of the first mode is not exactly zero because of discretization error.

Animation of the first mode reveals that one arm is able to rotate as a rigid body (without any deformation).

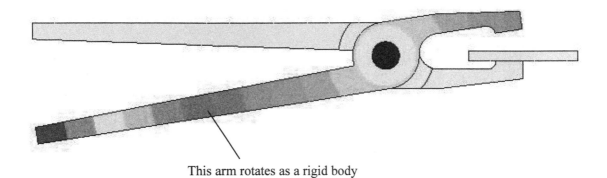

This arm rotates as a rigid body

Figure 10-4: Deformed plots of mode 1; your plot may show the rotating arm in the opposite extreme position.

Animate this shape with undeformed model superimposed on displacement plots.

Animation of the mode of vibration corresponding to the **Rigid Body Mode** demonstrates that the moving arm is disconnected from the flat plate. Review the model to find small gaps between the jaw and the plate. This is the cause of "insufficient restraints" in the model.

The first mode where deformation is present is mode 2; it's shown in Figure 10-5.

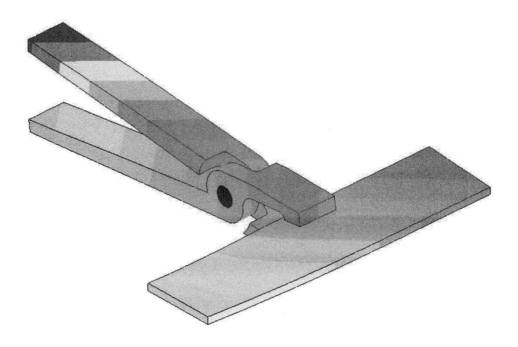

Figure 10-5: Deformed plot of mode 2.

The flat plate deforms as a cantilever beam.

You may want to switch to the *02 correct* assembly configuration and repeat the modal analysis (study *03 modal*). This configuration uses a thicker flat plate which touches both jaws. The touching faces are bonded by the automesher and the FEA model behaves as one solid body avoiding the problem of missing connectivity.

The problems with the PLIERS assembly have been identified using modal analysis which worked as a tool to detect RBMs. This in turn led us to the problem of disconnected faces. An alternative approach would be to use a **Static** study with the option **Use soft springs to stabilize model** selected.

The second example of using modal analysis as a diagnostic tool will be run on incorrectly supported model.

Open the part FLAT, which belongs to PLIERS assembly. We'll use it to demonstrate an application of modal analysis to identify a problem with restraints. The model comes with two studies defined: *01 static* and *02 modal*, both ready to run. An attempt to run *01 static* study produces the error message shown in Figure 10-6.

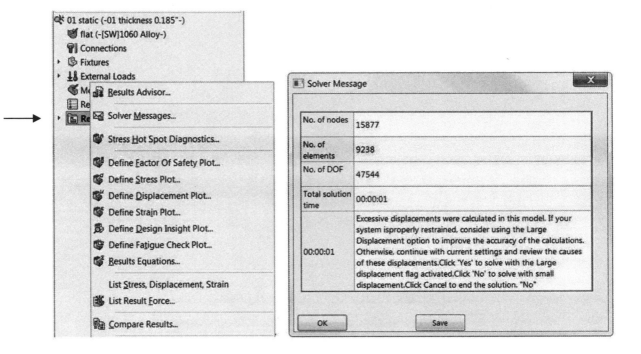

Figure 10-6: Solver error message in study *01 static*. Error message identifies a problem with excessive displacements.

This solver message is displayed if FFEPlus solver is used. Solution ends before any message is created if Direct Sparse Solver is used.

Selecting **Yes** in the Decision Solver Massage window (not shown here) fails the solution; selecting **No** produces clearly incorrect results with very large displacements.

The solution of study *02 modal* completes without an error. The **List Modes** table shows the first mode with a frequency of 0Hz (Figure 10-7).

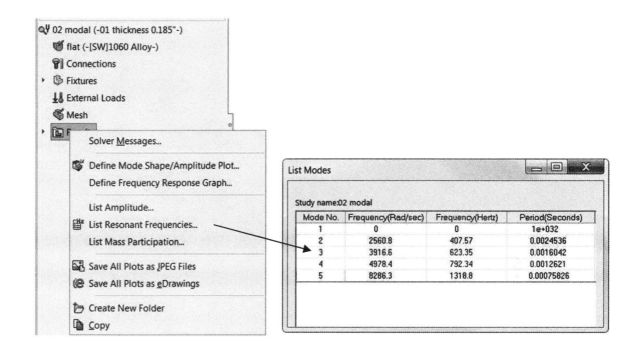

Figure 10-7: Results of *02 modal* study of part FLAT.

Mode 1 is a Rigid Body Mode with frequency 0Hz.

The modal frequencies may be listed by defining **Frequency Analysis Graph** (Figure 10-3) or using **List Resonant Frequencies** (Figure 10-7).

Animation of Mode 1 reveals the problem: the restraint is not fixed, even though it is called **Fixed**. The model has one RBM which is rotated about the edge where the **Fixed** restraint has been defined (Figure 10-8).

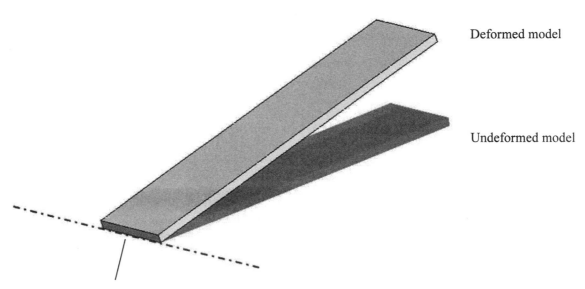

Deformed model

Undeformed model

The edge where **Fixed** restraint is defined becomes the axis of rotation

Figure 10-8: Result of modal analysis of FLAT part showing the shape of Mode 1 which is a Rigid Body Mode.

Even though we use the terms "Undeformed model" and "Deformed model," there is no deformation at all present in the model. The model rotates about the edge line where the Fixed restraints are defined.

The reason why an RBM is present in the model where the **Fixed** restraint has been applied is that solid elements have three degrees of freedom per node and these are all translations. Consequently, nodes of solid elements can't generate a moment reaction necessary to support the model and the model rotates about the line of support.

These are two simple examples of using modal analysis as a diagnostic tool. Both are based on the ability of modal analysis to identify rigid body motions. Structural analysis won't run with rigid body motions.

You may have noticed that we alternate between two terms: **Rigid Body Motion** and **Rigid Body Mode**. The **Rigid Body Motion** is a displacement without any deformation. The **Rigid Body Mode** is the way of identifying **Rigid Body Motion** by modal analysis which assigns to it frequency 0Hz.

Summary of studies completed

Model	Configuration	Study Name	Study Type
PLIERS.sldasm	01 error	01 static	Static
		02 modal	Frequency
	02 correct	03 modal	Frequency
FLAT.sldprt	01 thickness 0.185"	01 static	Static
		02 modal	Frequency

Figure 10-9: Names and types of studies completed in this chapter.

11: Time response and frequency response of discrete systems

Topics covered

- ❑ Time response
- ❑ Steady state harmonic response
- ❑ Frequency sweep
- ❑ Displacement base excitation
- ❑ Velocity base excitation
- ❑ Acceleration base excitation
- ❑ Resonance
- ❑ Modal damping

In the previous chapters, we studied different applications of modal analysis which investigated structures' propensity to vibrate but that was not a vibration analysis. In chapter 1 we have introduced the **Modal Superposition Method** which is an ubiquitous tool in Vibration Analysis. We'll be using it in all linear vibration problems.

Here is a short review of the **Modal Superposition Method**:

Modal Superposition Method

The **Modal Superposition Method** (MSM) represents a vibration response of a structure by using the superposition of responses that characterize Single Degree Of Freedom (SDOF) systems. The natural frequencies and directions of vibration of these SDOF systems correspond to the natural frequencies of the analyzed structure. The number of SDOFs contributing to a dynamic response is equal to the number of modes calculated by a pre-requisite modal (frequency) analysis.

How many modes should be calculated to represent a vibration response using the MSM? The first few modes are the most important, but the exact number of required modes is not known prior to analysis. Ideally, one should use a convergence process to demonstrate that increasing the number of modes past a certain number no longer significantly affects results.

Vibration analysis based on the MSM may be categorized into two major types: **Time Response** analysis and **Frequency Response** analysis. **Time Response** analysis is called **Modal Time History** in **SOLIDWORKS Simulation**; a **Frequency Response** analysis is called **Harmonic**. We'll be alternating between these terms just like we alternate between the terms **Modal** analysis and **Frequency** analysis.

Time Response (Modal Time History) analysis

In a **Time Response** analysis, the applied load is an explicit function of time. Damping is taken into consideration and the vibration equation appears in its full form:

$$[M]\ddot{d} + [C]\dot{d} + [K]d = F(t)$$

Where:

[M]	mass matrix
[C]	damping matrix
[K]	stiffness matrix
F(t)	vector of nodal loads (this vector is a function of time)
d	unknown vector of nodal displacements

A **Time Response** analysis requires the definition of a damping coefficient which is most often expressed as a percentage of critical damping. Readers are referred to (9) listed in Chapter 21 for numerical values of damping coefficients.

A **Time Response** analysis is used to model events of a short duration. A typical example would be an analysis of a structure's vibrations due to an impact load or acceleration applied to the base (called base excitation). Results of the **Time Response** analysis capture both the transient response and steady state vibration response.

Frequency Response (Harmonic) analysis

A **Harmonic** analysis assumes that the load is a function of frequency rather than being directly dependent on time as is the case of a **Time Response** analysis.

$$[M]\ddot{d} + [C]\dot{d} + [K]d = F(A\sin(\omega t) + B\cos(\omega t))$$

Where:

[M]	mass matrix
[C]	damping matrix
[K]	stiffness matrix
F(t)	vector of nodal loads (this vector is a function of frequency)
d	unknown vector of nodal displacements

A **Frequency Response** analysis models a structure's response to a forced excitation or base excitation (excitation applied to its supports) that is a harmonic function of time. It is assumed that the excitation frequency changes very slowly, hence the alternative name **Steady State Harmonic Response** is often used for this type of analysis. A **Frequency Response** analysis also uses the modal superposition method and requires that damping be defined, usually as a percentage of critical damping.

A typical application of a **Frequency Response** analysis is a simulation of a shaker table test discussed in Chapter 12.

In this chapter we introduce both **Time Response** and **Frequency Response** analyses using a simple discrete vibrating system. Open assembly model MDOF and review the three configurations *1DOF, 2DOF, and 3DOF* as shown in Figure 11-1. The same figure shows sensor locations in each configuration.

We'll use the *MDOF* assembly to run as many as ten studies. Because of this complexity we present the summary now rather than at the end of the chapter.

Summary of studies

All configurations and all studies are in the assembly model MDOF.sldasm.

Configuration	Study name	Study type	Description
1DOF	*01 1DOF modal*	Frequency	Modal analysis
2DOF	*02 2DOF modal*	Frequency	Modal analysis
3DOF	*03 3DOF modal*	Frequency	Modal analysis
1DOF	*04 1DOF disp fr*	Linear Dynamic Harmonic	Frequency Response analysis with displacement base excitation
1DOF	*05 1DOF vel fr*	Linear Dynamic Harmonic	Frequency Response analysis with velocity base excitation
1DOF	*06 1DOF acc fr*	Linear Dynamic Harmonic	Frequency Response analysis with acceleration base excitation
2DOF	*07 2DOF disp fr*	Linear Dynamic Harmonic	Frequency Response analysis with displacement base excitation
3DOF	*08 3DOF disp fr*	Linear Dynamic Harmonic	Frequency Response analysis with displacement base excitation
1DOF	*09 1DOF shock tr*	Linear Dynamic Modal Time History	Time Response analysis with shock load
1DOF	*10 1DOF initial cond tr*	Linear Dynamic Modal Time History	Time Response analysis with initial condition

Figure 11-3: MDOF assembly configurations and studies presented in this chapter.

We'll now discuss restraints, spring connectors and meshing which are common to all configurations and all studies.

All cubes have **Roller/Slider** restraints defined to four side faces (Figure 11-4).

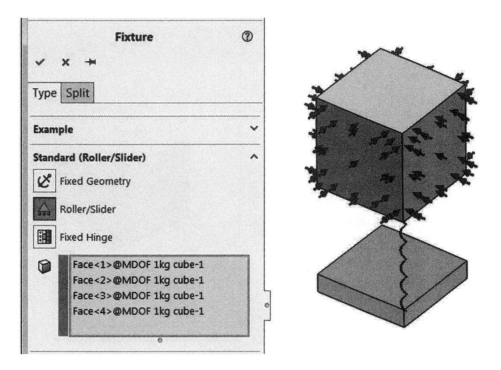

Figure 11-4: Roller/Slider restraints applied to four sides of each cube shown here for configuration *1DOF*.

Roller/Slider restraints may be applied only to two perpendicular faces; this will be sufficient to limit the motion to the vertical direction only.

Symbol of Spring Connector is shown; symbols of restraints applied to the base are not shown.

The base part has **Fixed** restraints applied; they may be applied to one face, to selected faces or to all faces.

All springs are modeled as spring connectors between two locations (Figure 11-5).

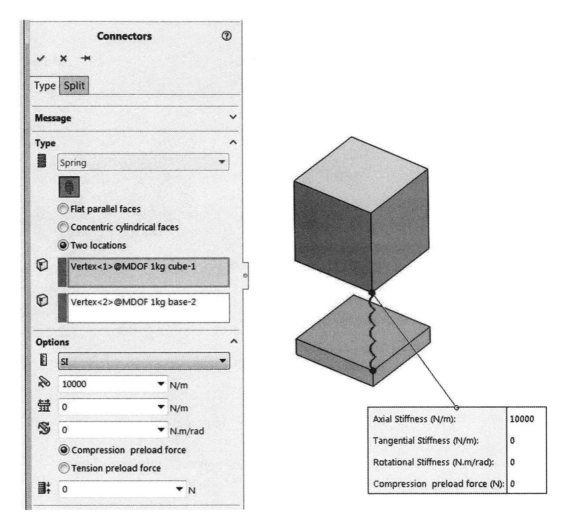

Figure 11-5: Spring connector shown for the model in configuration *1DOF*.

10000N/m is the linear stiffness.

We are modeling discrete systems. In these systems stiffness properties are fully defined by the springs and inertial properties are defined by the cubes. As we'll soon demonstrate, the cubes do not deform in these modes which are of interest to us. Therefore, the cubes and base may be meshed with a coarse mesh; use element size 20mm in all configurations.

Run **Frequency** studies on configurations *01 1DOF modal, 02 2DOF modal*, and *03 3DOF modal* to find the modes of vibration. You will find that the number of modes that are associated with the deformation of the springs corresponds to the number of degrees of freedom in each configuration. Remember, when we say "the number of degrees of freedom" we are referring to assembly components (cubes and base) as rigid bodies.

Results of modal analyses are summarized in Figure 11-6.

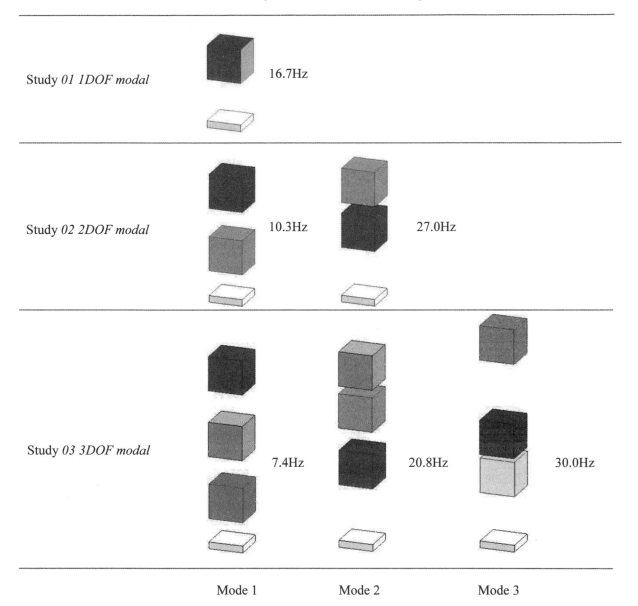

Mode 1   Mode 2   Mode 3

Figure 11-6: Summary of the results of modal analysis in all assembly configurations.

Configurations are arranged in rows, modes are arranged in columns.

You may want to review the shapes of higher modes to see that they are associated with deformation of cubes and do not correspond to the discrete system we analyze here.

Configuration 1DOF has an analytical solution $\omega = \sqrt{k/m} = 100$rad/s which is very close to the numerical solution 105rad/s or 16.7Hz.

From the results of the modal analysis presented in Figure 11-5, we learned that the range of frequencies, 0 – 30Hz, includes all natural frequencies of the system in all assembly configurations.

In the **Frequency Response** analysis that we are about to perform, we subject the base to oscillations with frequencies from zero (no motion) to frequencies much higher than the highest frequency. A frequency three times as high as the natural frequency is generally considered as a "much higher" frequency; therefore, the range of base excitation frequencies will be 0-90Hz. In the **Harmonic** analysis with base excitation, the base is subjected to harmonic motion as shown in Figure 11-7.

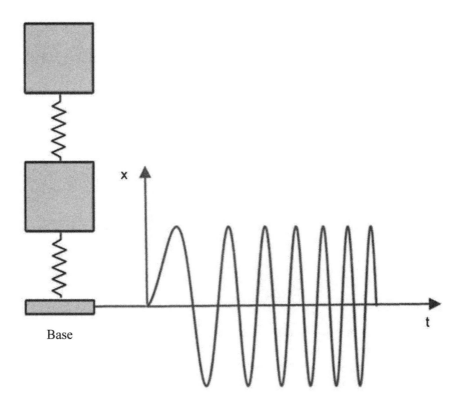

Figure 11-7: Harmonic base excitation applied to the base shown here for configuration *2DOF.*

The base performs harmonic motion with constant amplitude and increasing frequency. This excites vibration of two masses attached to the base.

Three different types of base excitation will be applied:

- Excitation with constant displacement amplitude
- Excitation with constant velocity amplitude
- Excitation with constant acceleration amplitude

Start in the *1 DOF* assembly configuration and create a **Harmonic** study titled *04 1DOF disp fr* as shown in Figure 11-8. This is the first study in the series of three **Harmonic** studies: *04 1DOF disp fr, 05 1DOF vel fr,* and *06 1DOF acc fr.*

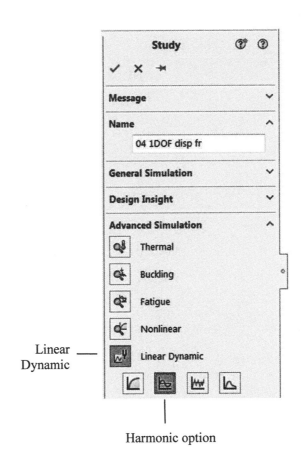

Linear Dynamic

Harmonic option

Figure 11-8: Definition of a Harmonic study.

Harmonic study is one of four options in Linear Dynamic study.

Define study properties as shown in Figure 11-9.

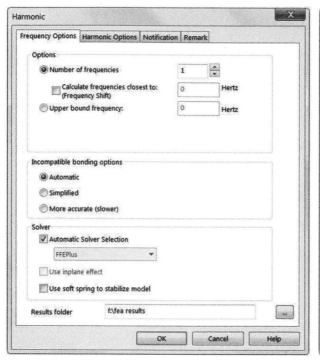 

Frequency Options   Harmonic Options

Figure 11-9: Definition of the Harmonic study.

The definition consists of specifying Frequency Options and Harmonic Options.

Like all **Linear Dynamic** studies, the **Harmonic** study is based on the **Modal Superposition Method.** The **Number of frequencies** specified in **Frequency Options** gives the number of modes to be used to find the vibration response. In this exercise we analyze a discrete vibration system with one degree of freedom which had one mode of vibration. For this reason the **Number of frequencies** is 1. The **Operating frequency limits** defined in **Harmonic Options** specifies the range of frequencies to which the model will be subjected. The vibration response will be found as function of the excitation frequency slowly changing from 0 to 90Hz.

Define a **Fixed** restraint to base and **Roller/Slider** restraints the same as in the previously completed **Frequency** analyses.

Define **Base Excitation** as shown in Figure 11-10.

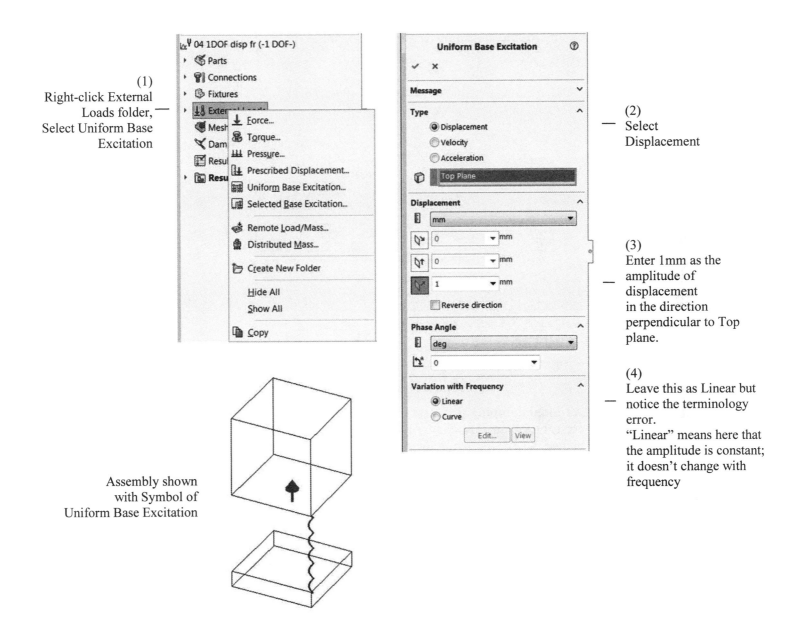

(1)
Right-click External Loads folder, Select Uniform Base Excitation

(2)
Select Displacement

(3)
Enter 1mm as the amplitude of displacement in the direction perpendicular to Top plane.

(4)
Leave this as Linear but notice the terminology error.
"Linear" means here that the amplitude is constant; it doesn't change with frequency

Assembly shown with Symbol of Uniform Base Excitation

Figure 11-10: Definition of Uniform Base Excitation.

The wireframe display is used to clearly show the symbol of Uniform Base Excitation. The symbols of the Roller/Slider restraints applied to all four sides of the cube and symbols for the Fixed restraints applied to all faces of the base are not shown in this illustration.

Uniform Base Excitation means that the base excitation is applied through all restraints active in the specified direction. In our case it means that it is applied to the base but not to the cube, because the cube does not have any restraints in the direction of excitation. **Variation with Frequency** is defined as **Linear** but this should say **Constant**; the amplitude doesn't change with the excitation frequency.

Apply **Modal Damping** as shown in Figure 11-11.

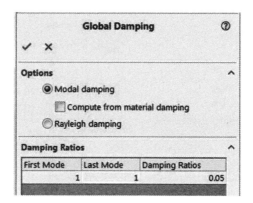

Figure 11-11: A Modal Damping of 5% is defined for the only mode present in this discrete vibrating system.

Modal damping may be defined directly or may be computed from material damping. Here, we use the first method.

A **Modal damping** of 5% means that damping in the system equals 5% of critical damping. Oscillatory motion is not possible for critical damping or higher. This definition of damping doesn't specify which model component is responsible for dissipating energy. This is why the window in Figure 11-11 is called **Global Damping**.

Define **Results Options** as shown in Figure 11-12:

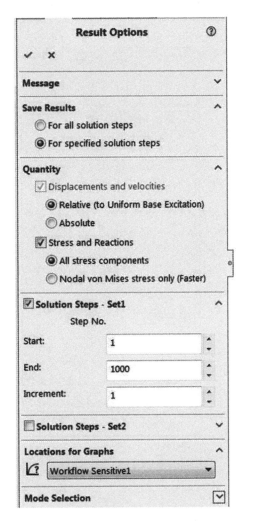
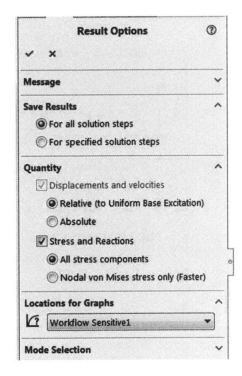

Save Results:
For specified solution steps

Save Results:
For all solution steps

Figure 11-12: Result Options definition.

The left window defines which steps will be used for creating Response Graphs. Here, all steps will be used. The number referring to the last step may be defined as higher than the actual number of preformed steps. The end step is defined as step 1000 which is the highest number allowed by the system.

The right window uses all steps. Using the above settings, Save Results for specified solution steps or Save Results for all solution steps produces the same results.

Obtain the solution and define a **Response Graph** as shown in Figure 11-13.

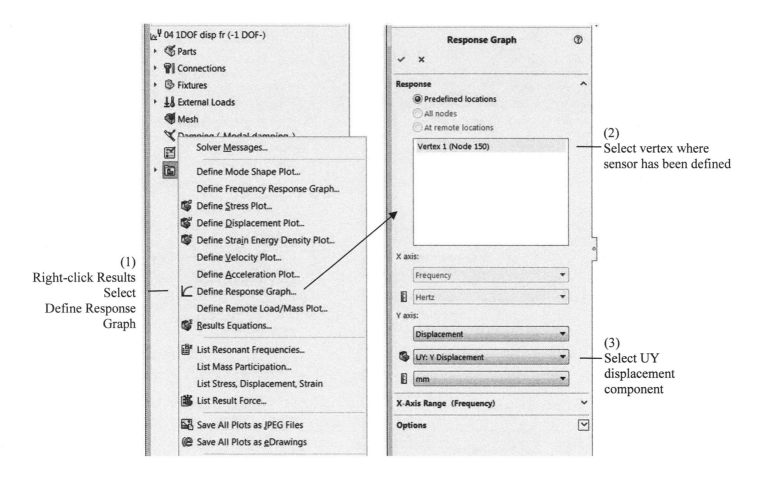

Figure 11-13: Response Graph definition.

Displacement component UY is in the direction of the base excitation.

Before analyzing results of study *04 1DOF disp fr*, we'll define and run two more **Harmonic** studies, then we'll analyze all the results together. Copy study *04 1DOF disp fr* into *05 1DOF vel fr*, then copy it into *06 1DOF acc fr*. Edit the two new studies as shown in Figure 11-14; the only difference between these three studies is in the base excitation definition.

Study *04 1DOF disp fr*
Displacement excitation
amplitude 0.001m

Study *05 1DOF vel fr*
Velocity excitation
amplitude 0.1m/s

Study *06 1DOF acc fr*
Acceleration excitation
amplitude 10m/s²

Figure 11-14: Base excitation definition in three Harmonic studies with base excitation.

The left window is a repetition of Figure 11-10.

The results are summarized in Figure 11-15.

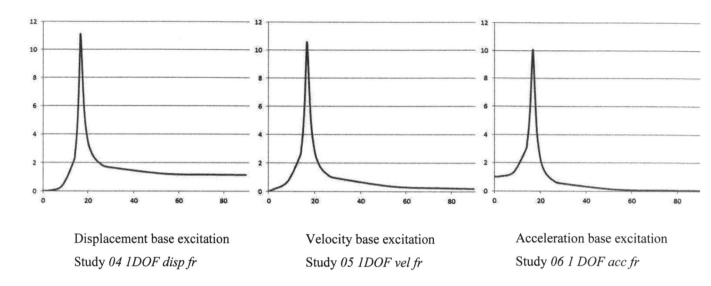

Displacement base excitation

Study *04 1DOF disp fr*

Velocity base excitation

Study *05 1DOF vel fr*

Acceleration base excitation

Study *06 1 DOF acc fr*

Figure 11-15: UY displacement amplitude as a function of the excitation frequency.

Unit of abscissa is Hz, unit of ordinate is mm. These graphs have been formatted in Excel.

UY displacement amplitude response graphs presented in Figure 11-15 all show one peak that corresponds to 16Hz (100rad/s) which is the natural frequency of the system in the *1DOF* configuration. For low damping, as is the case in this exercise, the amplitude of displacement reaches a maximum when the excitation frequency equals the natural frequency; this is called resonance.

Let's explain why in this exercise the displacement amplitude at resonance is the same for all three types of base excitation. Displacement, velocity, and acceleration in a harmonic motion are:

$$x = Asin(\omega t)$$

$$\dot{x} = A\omega cos(\omega t)$$

$$\ddot{x} = -A\omega^2 sin(\omega t)$$

Displacement amplitude is proportional to the base excitation, be it displacement, velocity, or acceleration excitation. Considering that $\omega = 100\,rad/s$ is the frequency of excitation at resonance, the numerical value of acceleration amplitude is 100 times higher than the velocity amplitude and 10000 higher than displacement amplitude. Our definitions of acceleration base excitation (10m/s²), velocity base excitation (0.1m/s) and displacement base excitation (0.001m) follow the same ratio. Therefore, the displacement amplitude at resonance is the same for all three types of base excitation.

Examination of the response graphs in Figure 11-15 reveals an important property of displacement base excitation: the amplitude of vibration remains constant for frequencies much higher than the resonance.

We'll now conduct the analysis of the assembly in *2DOF* and *3DOF* configurations, limiting it to displacement base excitation with amplitude of 1mm and a frequency sweep of 0-90Hz. In the properties of study *07 2DOFdisp fr* define 2 modes and in the properties of study *08 3DOF disp fr* define 3 modes in **Frequency Options**.

Define a **Base Excitation** as shown in Figure 11-10. Define **Modal Damping** for all modes present in each study. Spring connectors, restraints and the mesh remain the same.

Study *07 2DOFdisp fr* has two sensors; study *08 3DOFdisp fr* has three sensors; define **Results Options** for all sensors present in each study. Obtain solutions and define UY **Response Graphs**.

The frequency responses of UY displacement amplitude for all three studies are summarized in Figure 11-16.

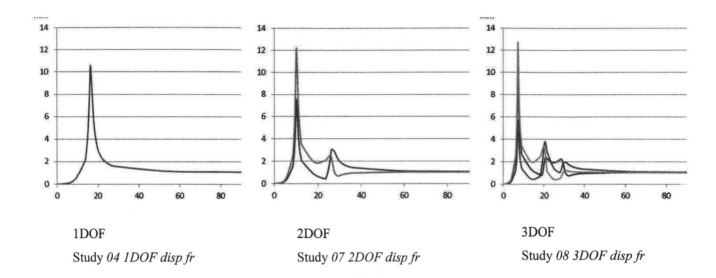

1DOF

Study *04 1DOF disp fr*

2DOF

Study *07 2DOF disp fr*

3DOF

Study *08 3DOF disp fr*

Figure 11-16: UY displacement amplitude as a function of the excitation frequency for the three assembly configurations.

Unit of abscissa is Hz, unit of ordinate is mm. These graphs have been formatted in Excel. The graph on the left has already been shown in Figure 11-15.

The number of peaks in these three frequency sweeps illustrated in Figure 11-15 equals the number of modes of vibration present in each study. The first mode dominates the displacement amplitude response in all cases. Please review the spreadsheet MDOF.xlsx for more details impossible to show in these black and white illustrations.

We now move to a **Time Response** analysis using the same *MDOF* assembly in configuration *1DOF*. Make sure the assembly is in *1DOF* configurations and create a study titled *09 1DOF shock tr* as shown in Figure 11-17.

Linear
Dynamic ———

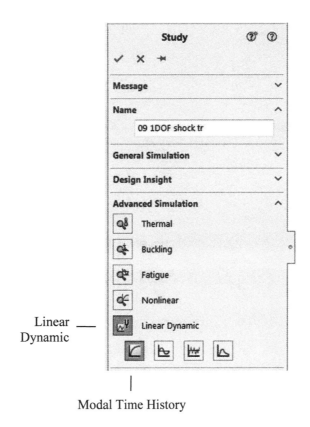

Modal Time History

Figure 11-17: Definition of a Time Response study called Modal Time History in SOLIDWORKS Simulation.

Modal Time History is one of the four options in Linear Dynamic study. Modal Time History analysis is also called Time Response analysis.

A vibration response in a **Time Response** study is a function of time. In the study properties we must specify the time duration of the analysis and the time step. The duration of analysis will be 10 times the period of the mode of vibration with 10 time steps per oscillation. Remembering that the natural frequency is 100rad/s, the period of vibration is:

$$T = \frac{2\,\Pi}{\omega} = 0.063s$$

The duration of analysis is 0.63s and the time step 0.0063s. Define the properties of the **Modal Time History** study as shown in Figure 11-18.

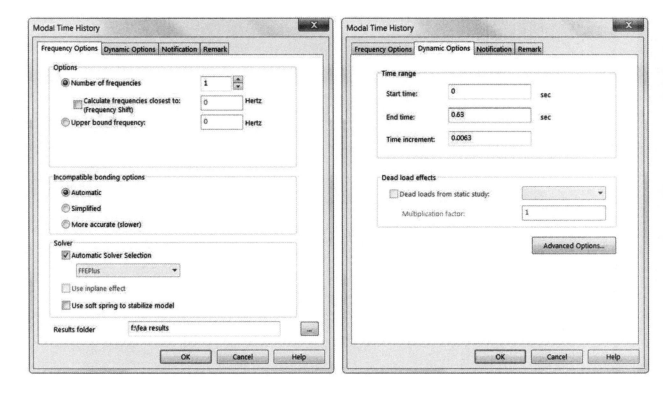

Frequency Options   Dynamic Options

Figure 11-18: Properties of the Time response study.

Given time duration of 0.63s and a time step of 0.0063, the analysis will complete in 100 steps.

Use the same restraints, **Results Options**, and mesh size as in study *04 1DOF disp fr*.

Damping in a **Modal Time History** analysis may be defined in two different ways: as **Modal Damping**, the same as in a **Harmonic** study, or as linear damping in a **Spring-Damper** connector.

We'll work with the same 5% modal damping as before but we'll express it as linear damping in the definition of a **Spring-Damper** connector. The critical damping in a Single Degree of Freedom oscillator is:

$$c_{cr} = 2\sqrt{km} = 200\,\frac{Ns}{m} \qquad k = 10000\,\frac{N}{m} \quad m = 1kg$$

$$c = 0.05c_{cr} = 10\,\frac{Ns}{m}$$

Therefore, to define damping as 5% of critical damping we can either enter 10Ns/m in the **Spring-Damper Connector** window, or as 0.05 in the **Global Damping** window (Figure 11-19).

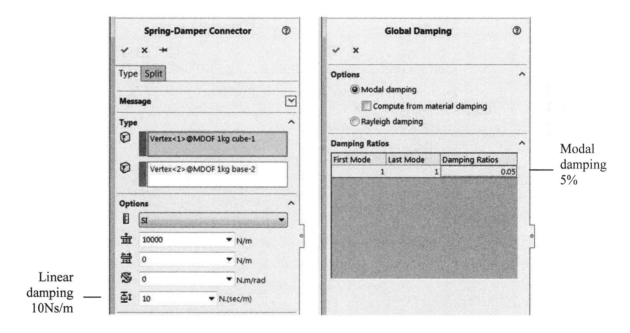

Linear damping definition   Modal damping definition

<u>Figure 11-19: Damping can be defined explicitly as a linear damping (left) or as a fraction of critical damping (right). The entries in both windows define the same damping.</u>

Define damping as a linear damping. If you define damping using both methods at the same time, the resultant damping will be 10% modal.

A **Time Response** analysis requires a load defined as a function of time. Here, we want to model the system response to a shock load, which is a load of short duration. In our example the load duration is 0.01s, the maximum load magnitude is 1000N, and the shape of the load time history curve is half a sine curve. (Figure 11-20).

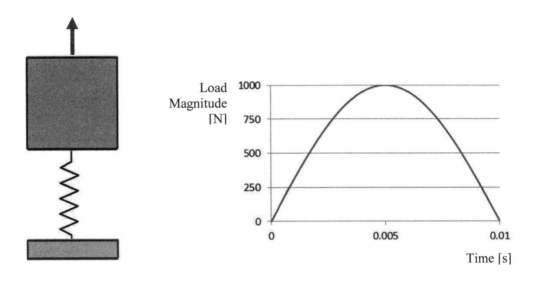

Figure 11-20: Shock load applied in the direction of motion of 1DOF model.

Load is applied to the top face; it disappears after 0.01s.

To define this load, follow the steps in Figure 11-21.

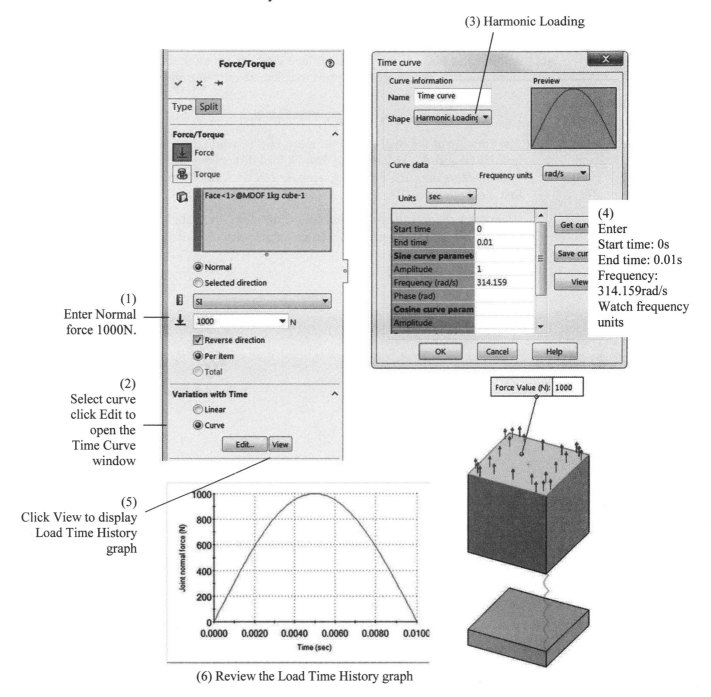

(6) Review the Load Time History graph

Figure 11-21: Defining load as a function of time. The force takes 0.005s to reach the maximum of 1000N, then another 0.005s to drop back to zero.

Select Variation with Time as Curve (2), and click Edit to open the Time curve definition window. Define the shape as harmonic loading (3) and enter values as shown (4). Click View (5) to examine the Load time history curve (6).

Notice that neither the entry in the **Force** window (here 1000N) nor the values defining the **Time curve** define the load time history on their own. The corresponding values are multiplied to calculate the force magnitude as a function of time.

Define **Results Options** as shown in Figure 11-12.

The duration of the load is 0.01s (Figure 11-20), while the duration of analysis is 0.63s (Figure 11-18) meaning that the load disappears after 0.01s and the system enters into free vibration. Run *09 1DOF tr shock* study and observe that it completes in 100 steps.

Right-click the *Results* folder and follow the steps illustrated in Figure 11-22 to create a graph showing displacement in the sensor location as a function of time.

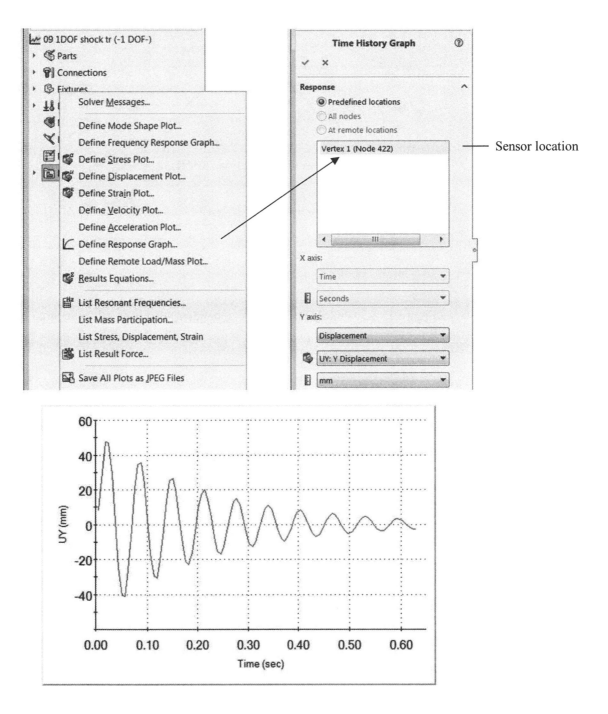

Figure 11-22: Displacement of the 1kg mass during the first 0.63s after load application.

Notice that after 0.01s, excitation becomes zero and the mass performs free damped oscillations.

There is one more study left to run. Copy the last study *09 1DOF shock tr* into *10 1DOF initial cond tr*. These two studies are identical except for the method used to excite vibration. In study *09 1DOF shock tr* we used a load of very short duration and called it a Shock Load. This time dependent load acting over 0.01s generates an impulse load of 6.31Ns (Figure 11-23).

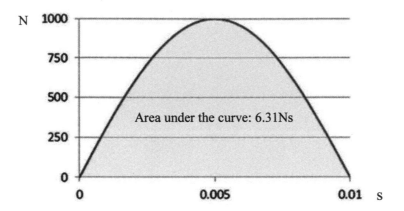

Figure 11-23: Impulse of the time dependent load is equal to the area under the curve.

See MDOF IMPULSE.xlsx

Considering that:

$$mv_0 = Ft \quad \text{where } m = 1kg \quad Ft = 6.31Ns$$

we find the initial velocity of the 1kg mass that gives an equivalent excitation as the above time dependent load:

$$v_0 = \frac{Ft}{m} = 6.31\frac{m}{s}$$

In study *10 1DOF initial cond tr* we set the system in motion by using an initial condition rather than a time dependent load.

Delete the load that was copied from study *09 1DOF shock tr* and define an initial condition as shown in Figure 11-24.

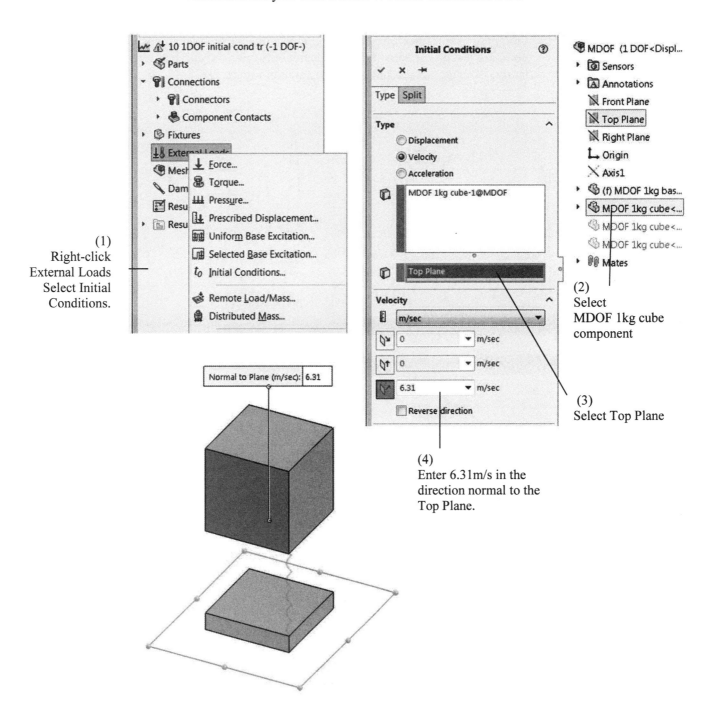

(1)
Right-click
External Loads
Select Initial
Conditions.

(2)
Select
MDOF 1kg cube
component

(3)
Select Top Plane

(4)
Enter 6.31m/s in the
direction normal to the
Top Plane.

Figure 11-24: Definition of the initial velocity.

The initial velocity definition applies to the entire cube.

Obtain the solution and review the UY displacement response graph and compare it with the same UY displacement response graph from the previous study. Both graphs are shown side by side in Figure 11-25.

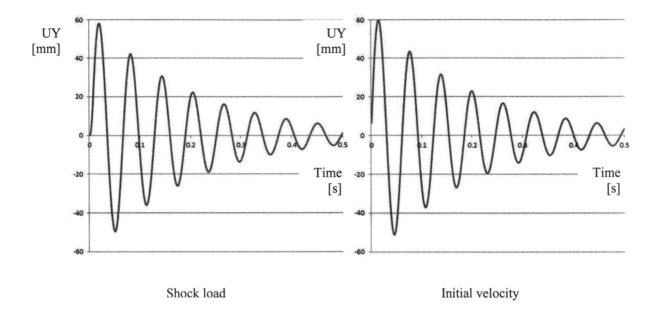

Shock load   Initial velocity

Figure 11-25: UY displacement response to the shock load and to the initial velocity.

Both graphs are identical except for the first 0.01s when the shock load is active.

Results shown in Figure 11-24 show that in a single degree of freedom system the effect of a shock load (load of a very short duration) may be modeled by an equivalent initial velocity.

You may want to extend this exercise even further and investigate the effect of different damping on the vibration response. This may be easily done using, for example, study *06 1DOF acc fr*. This study was run with a modal damping 5%. Re-run the same study with damping 2% and 10% and summarize the UY displacement responses in one graph as shown in Figure 11-26.

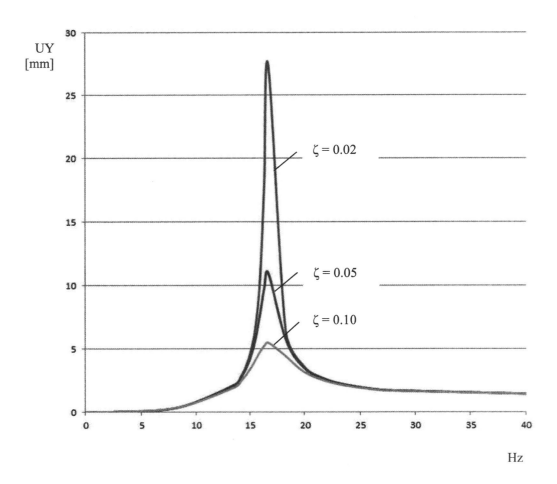

<u>Figure 11-26: The amplitude of vibration as a function of excitation frequency for different modal damping values.</u>

Damping strongly affects the amplitude for excitation frequencies close to the resonant frequency. It has no effect for excitation frequencies much lower or much higher than the resonant frequency.

Remember that the amplitude of vibration is always measured from the neutral position, not between negative and positive peaks.

Harmonic base excitation may be realized experimentally by placing the tested structure on a shaker table which oscillates with displacement amplitude A and frequency ω (Figure 12-1).

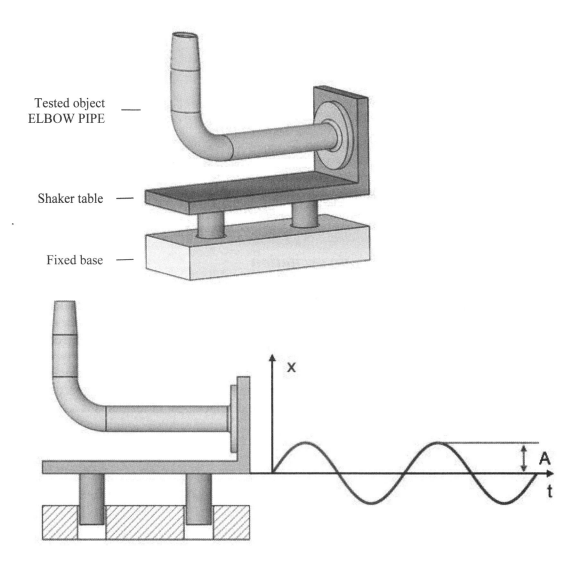

Tested object
ELBOW PIPE

Shaker table

Fixed base

Figure 12-1: Schematic representation of a shaker table.

The shaker table oscillates relative to the base. It moves up and down with displacement amplitude A and frequency ω.

In the lower illustration cross-section view is used to show the base; the table is shown in the neutral (mid-stroke) position. The coordinate system is fixed to the base, meaning that displacement is measured relative to the fixed base.

The model shown in Figure 12-1 is the SHAKER assembly. It is for show only and won't be analyzed; a component of this assembly, ELBOW PIPE, will be analyzed.

A shaker test is often conducted in such a way that the displacement amplitude remains constant while the frequency of excitation changes within a certain range. This is done to investigate the response of a tested object to excitation with different frequencies. The test is called a **Frequency Sweep** and will be simulated in this chapter by subjecting the part model ELBOW PIPE to base excitation in the frequency range of 0 - 400Hz.

We use the frequency range of 0 - 400Hz because test data demonstrate that this is the range of excitation frequencies the part will see in service. A very important assumption has to be made: the rate of change of the excitation frequency is slow. Therefore, the vibration response is a steady state response; transient response is not modeled. As we already know from Chapter 11, in **SOLIDWORKS Simulation** the analysis we need to conduct is called the **Linear Dynamic** study with a **Harmonic** option. Vibration analysis textbooks call this a **Steady State Harmonic Response**.

In preparation for the **Steady State Harmonic Response** analysis, we have to run two preparatory analyses.

First we'll run a static study to identify the location of stress concentrations and to design a mesh that will correctly model them. Using information from that static study we'll know where to locate stress sensors.

Next, we'll run a modal analysis to see how many modes of vibration are in the 0 - 400Hz frequency range. This way we'll know how many modes have to be considered in the vibration response analysis based on the **Modal Superposition Method**.

Open part model ELBOW PIPE and create a **Static** study *Gravity*. Apply a gravity load of 9.81m/s² and a fixed restraint as shown in Figure 12-2. The same restraint will be used in all studies in this chapter.

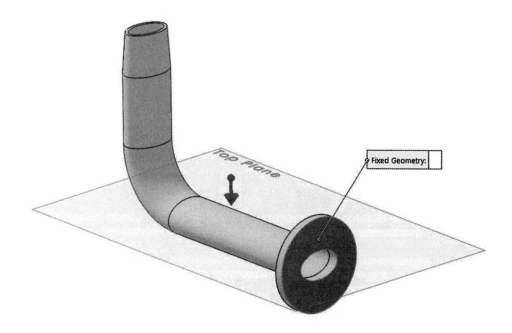

Figure 12-2: Load in the static study is the gravity load.

Gravity load 9.81m/s² is normal to the Top reference plane.

Fixed restraint is applied to the flange face.

This load is not supposed to model any real life loading conditions during the frequency sweep; weight has a negligible effect on stresses. However, being a volume load, gravity load is qualitatively close to the loading that the model will experience during the **Frequency Sweep**. We need it only to locate stress concentrations and to design the mesh. Absolute magnitudes of stress do not matter.

Define a mesh control on the fillet near the flange as shown in Figure 12-3.

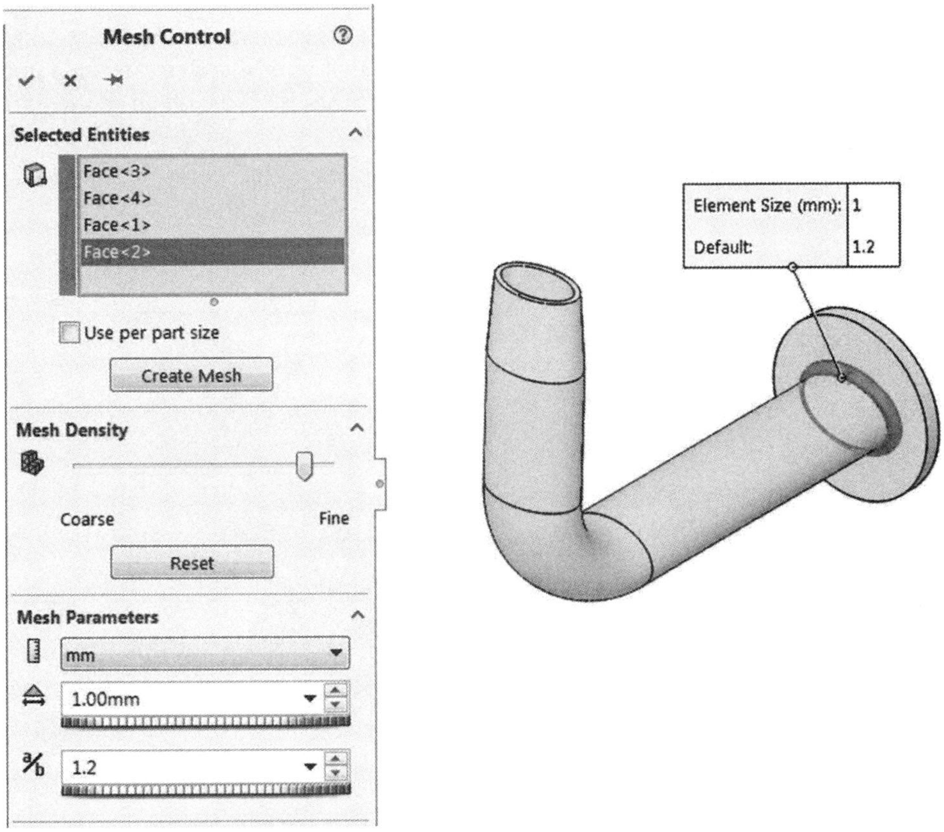

Figure 12-3: Mesh control to the fillet; notice that the fillet face is divided into four faces by a split line.

Use a 1mm element size on the controlled entities and the ratio of transition, a/b, equal to 1.2.

Mesh the model with a global element size of 5mm and obtain the static solution. You may use a less refined mesh to reduce solution times in this and following studies.

Review **P1 stress** results and confirm the location of the stress concentrations as shown in Figure 12-4. Remember that von Mises stress is not applicable to brittle material **Gray Cast Iron** which has been assigned to the model.

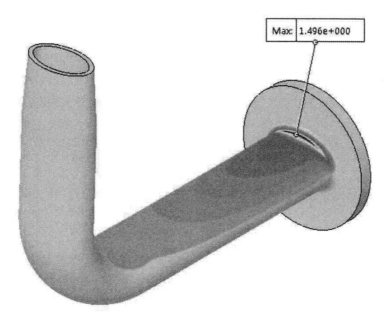

Max: 1.496e+000

Figure 12-4: Location of the maximum stress.

This location of stress concentration coincides with the crossing of two split lines placed there to mark the location of a stress sensor.

Based on the results of static stress analysis, we define two sensors in the location shown in Figure 12-5. Sensor 1 will be used to create **Displacement Response** graphs. Sensor 2 will be used to create **Stress Response** graphs.

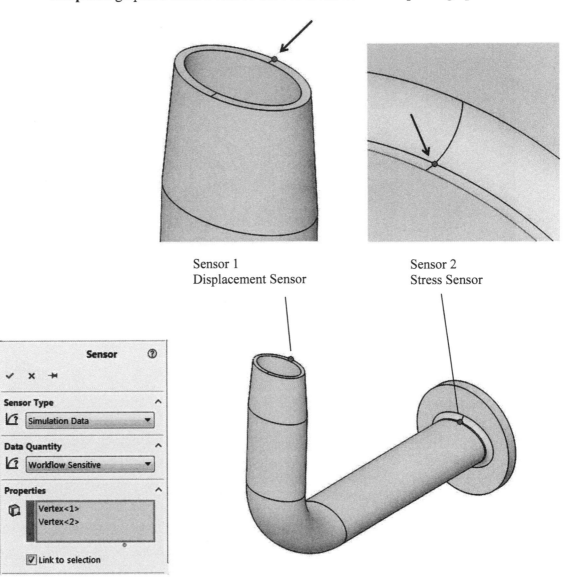

Sensor 1
Displacement Sensor

Sensor 2
Stress Sensor

Figure 12-5: Sensor locations: two locations are defined in one Sensor window; both sensors use vertices defined by split lines.

This is for information only; both sensors are already defined in the SOLIDWORKS model.

Create a **Frequency** study titled *Modal* with properties shown in Figure 12-6.

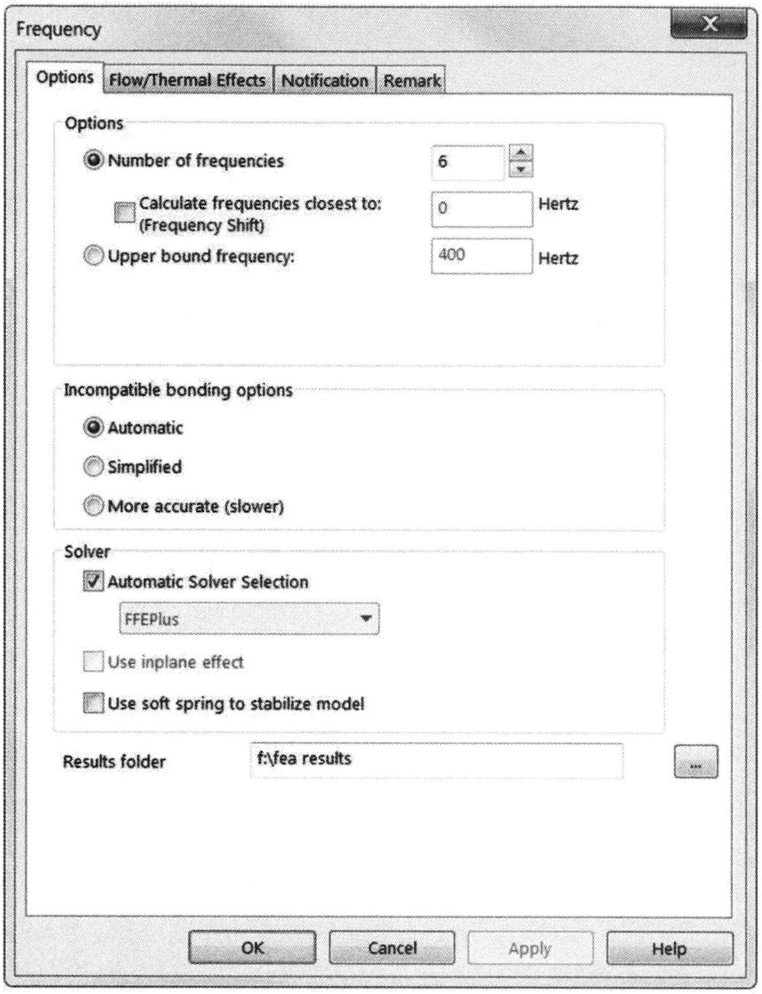

Figure 12-6: Properties of the frequency study; six modes requested.

We could have also specified the Upper bound frequency higher than 400Hz; e.g. 800Hz.

We try six modes to see if six modes cover the range of 0-400Hz and above 400Hz. Those modes within the range 0-400Hz and slightly above are important in vibration response of ELBOW PIPE.

Obtain the modal solution and list the modes in the 0-400Hz range as shown in Figure 12-7. Remember that you may also run a modal analysis within the **Harmonic** study as explained in Chapter 11.

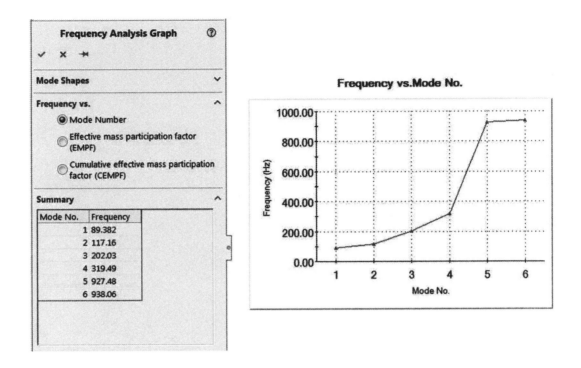

Figure 12-7: Results of modal analysis show four modes within the range of 0-400Hz.

Mode 5(927Hz) and mode 6 (938Hz) are far above the frequency range 0-400Hz.

Review modal shapes shown in Figure 12-8:

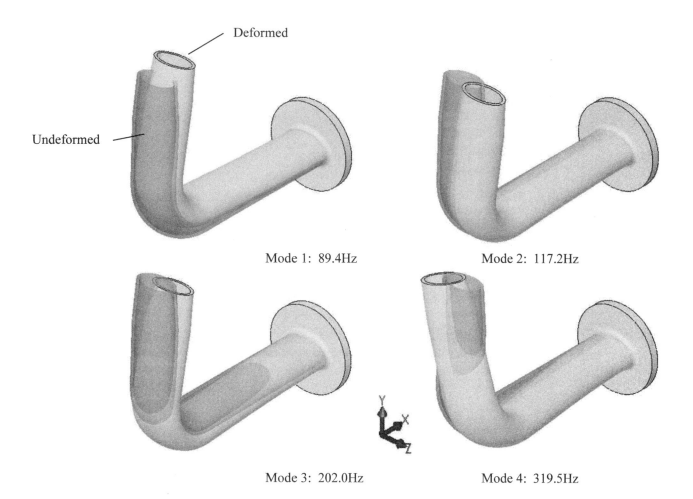

Mode 1: 89.4Hz   Mode 2: 117.2Hz

Mode 3: 202.0Hz   Mode 4: 319.5Hz

Figure 12-8: Modal shapes shown with the undeformed model superimposed on the deformed model.

All modes are shown in the same view.

As shown in Figure 12-8, vibration in modes 1 and 3 take place in the XY plane; vibration in modes 2 and 4 takes place in YZ plane. The base excitation is applied in the Y direction; therefore, only mode 1 and mode 3 will be excited in the **Frequency Sweep** simulation.

In real life the direction of excitation would never be perfectly aligned with one plane and all modes would be excited to some extent.

We will simulate a frequency sweep with a shaker table using **Harmonic** study *Frequency sweep* with properties shown in Figure 12-9.

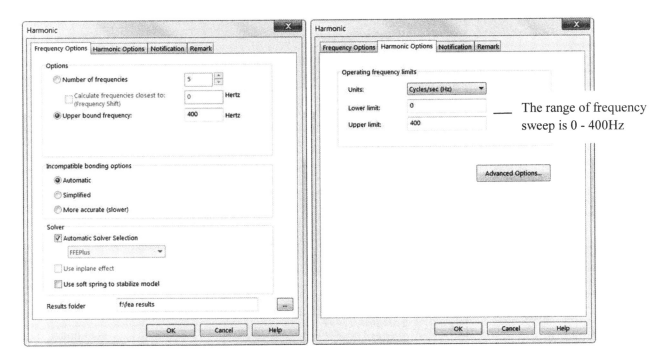

Frequency options   Harmonic Options

Figure 12-9: Frequency Options and Harmonic Options in Harmonic study simulating a Frequency Sweep.

Frequency Options define how many modes will be used in the Modal Superposition method.

Harmonic Options define the range of frequencies of Frequency Sweep.

Define **Results Options** as shown in Figure 12-12.

For all
solution steps ——

Relative ——

All stress
components ——

All Tracked
Data Sensors ——

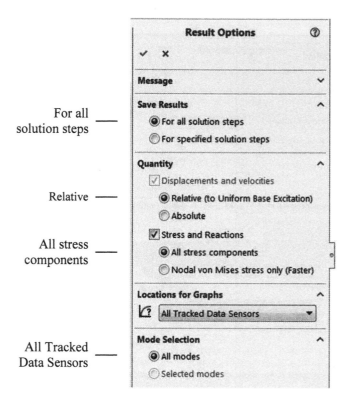

Figure 12-12: Results Options definition in *Frequency sweep* study.

Relative means that results are measured relative to the oscillating table.

Construct **Response Graphs** of resultant displacement amplitude recorded by the **Displacement Sensor** and P1 stress amplitude recorded by the **Stress Sensor**.

Definition of the **Response Graph** for resultant displacement amplitude is shown in Figure 12-13.

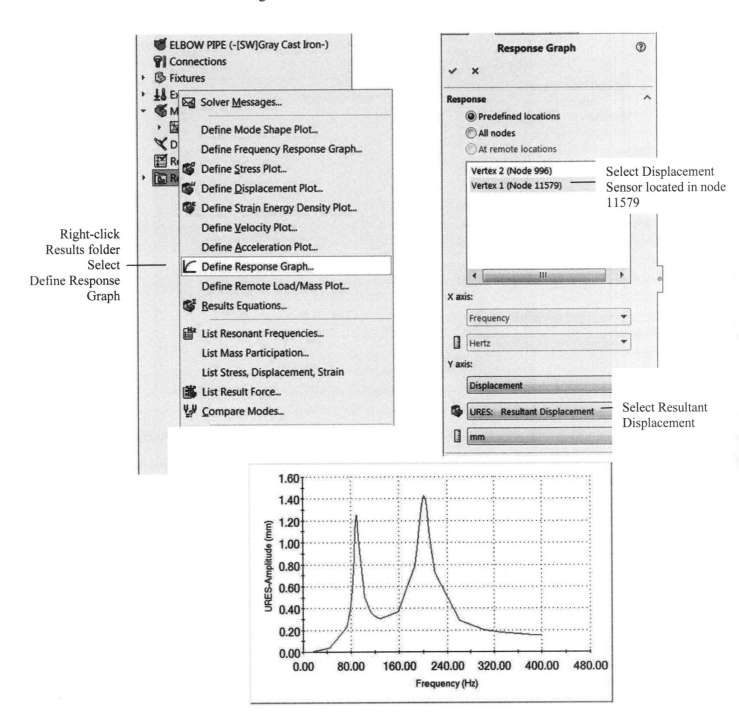

Figure 12-13: Definition of the Response Graph shown here for resultant displacement.

Select the vertex where the required sensor is located.

Automatically created graphs may be saved as a csv file for processing in a spreadsheet (Figure 12-14).

Click File
Select
Save As csv

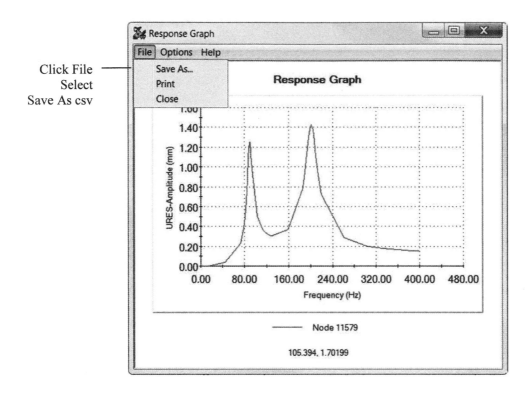

Figure 12-14: Saving the Response Graph for processing in another program.

Select csv format to export data to Excel. Selection of the csv data format is not shown in this illustration.

The graph in Figure 12-15 shows UY displacement magnitude as a function of frequency; it has been formatted in Excel.

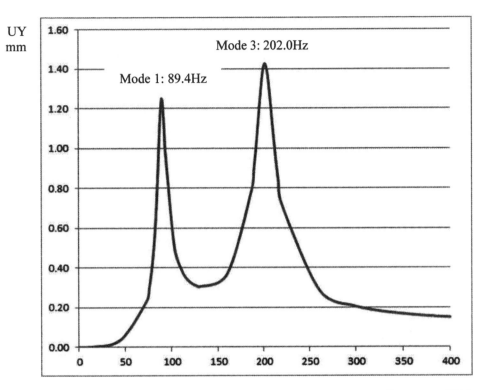

Figure 12-15: Displacement amplitude as a function of excitation frequency. The excitation frequency changes within the range of the frequency sweep (0-400Hz).

The excitation is in the XY plane. Therefore, modes 2 and 4, which have shapes moving in the orthogonal plane YZ, are not excited even though their frequencies are within the range of the **Frequency Sweep**.

Review the Response Graph of the UZ displacement component and notice that UZ displacement is practically zero. The only reason some displacement is shown is due to discretization error which makes stiffness not exactly symmetric about the XY plane.

The graph in Figure 12-16 shows P1 stress as a function of frequency. It has also been formatted in Excel.

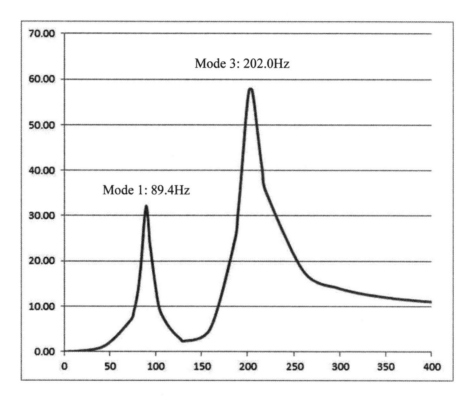

Figure 12-16: P1 stress amplitude in the indicated location as a function of excitation frequency. The excitation frequency changes within the range of the frequency sweep (0-400Hz).

The excitation frequency equal to the third modal frequency 199Hz produces P1 stress equal to 47MPa.

Excitation with frequencies equal to or very close to the modal frequency and in the direction of the modal shape is called **Resonance**. At resonance, inertial forces cancel with stiffness forces and the vibration response is controlled only by damping. This causes a very significant increase of displacement and stress amplitudes at resonant frequencies.

Summary of studies completed

Model	Configuration	Study Name	Study Type
ELBOW PIPE.sldprt	*Default*	*Gravity*	Static
		Modal	Frequency
		Frequency sweep	Harmonic

Figure 12-17: Names and types of studies completed in this chapter.

The drum has a dynamic imbalance of 5kg at a radius of 100mm. The centrifugal imbalance force F is a function of angular velocity, ω:

$$F = me\omega^2$$

$$m = 5kg \ - \ imbalanced \ mass$$

$$e = 0.1m \ - \ eccentricity$$

$$\omega \ \ rad/s - \ angular \ velocity$$

The operating speed of the centrifuge is 2000RPM which corresponds to a frequency of 33.3Hz. Our objective is to find the amplitude of vibration of the centrifuge body as a function of the angular velocity when the machine slowly reaches the operating speed. This is **a Steady State Harmonic Response** problem requiring a **Harmonic** study.

Create a **Harmonic** study titled *Centrifuge* and define restraints as shown in Figure 13-2.

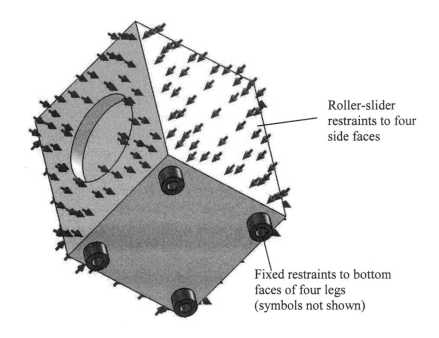

Roller-slider restraints to four side faces

Fixed restraints to bottom faces of four legs (symbols not shown)

Figure 13-2: Restraints applied to the assembly model allow for vertical movement of the body accompanied by deformation of the rubber mounts.

Roller-slider restraints applied to two perpendicular faces would suffice.

Symbols of Fixed restraints defined on the bottom faces of the four rubber mounts are not shown.

Even though geometry is very simple, meshing requires some planning. If the entire model is meshed with a default element size, rubber legs will be meshed with highly distorted elements; consequently leg stiffness will be modeled incorrectly; it will be too high. An incorrect mesh is shown in Figure 13-3.

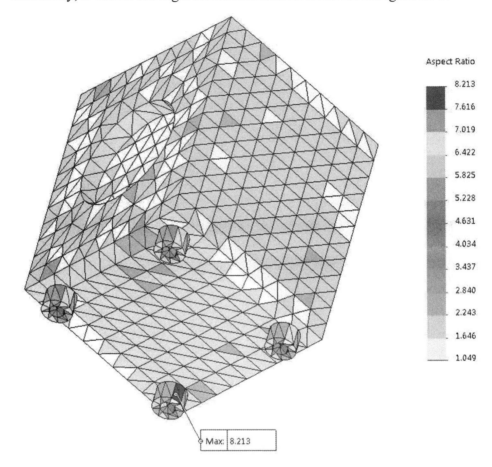

Figure 13-3: Incorrect mesh created using the default element size.

Aspect Ratio plot highlights highly distorted elements on all four legs.

The correct mesh must have small elements meshing legs while the centrifuge body may be meshed with large elements. This may be accomplished by defining **Mesh Control** on legs which are components of the assembly. Additionally, **Incompatible Mesh** in **Component Contact** options may be specified to avoid unnecessarily small elements in the centrifuge body close to legs. Mesh design is shown in Figure 13-4.

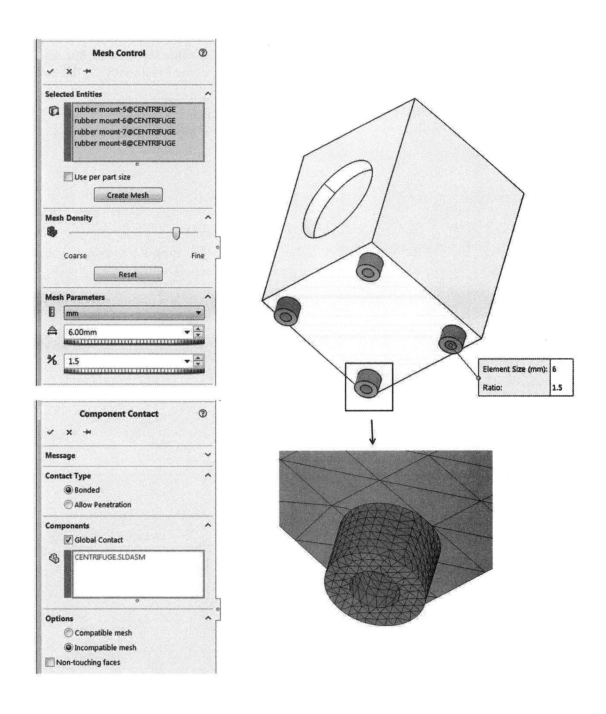

Figure 13-4: Correct mesh created with small elements meshing rubber legs and large elements meshing the centrifuge body.

Examine the detail and notice that nodes in rubber legs do not match the nodes in the centrifuge body because Incompatible mesh has been specified in Component Contact.

Before running the study, we'll run a modal analysis to investigate frequencies in the range of 0-2000RPM. Run a modal analysis from inside the **Harmonic** study and review the frequencies of the first five modes. Specify five modes instead of the default fifteen modes in the **Frequency Options** of the **Harmonic** study.

Results of the modal analysis are shown in Figure 13-5.

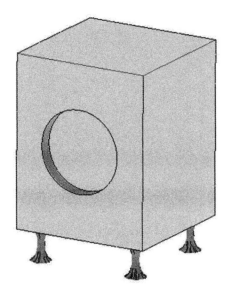

Mode 1 with frequency of 8.3Hz: Deformation is limited to the rubber legs. The centrifuge performs oscillations in the vertical direction.

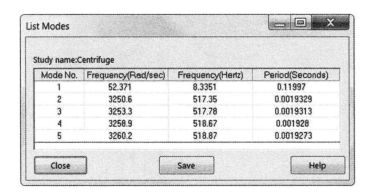

List Modes

Study name:Centrifuge

Mode No.	Frequency(Rad/sec)	Frequency(Hertz)	Period(Seconds)
1	52.371	8.3351	0.11997
2	3250.6	517.35	0.0019329
3	3253.3	517.78	0.0019313
4	3258.9	518.67	0.001928
5	3260.2	518.87	0.0019273

Only Mode 1 is in the range of the operating frequencies 0-2000RPM (0 – 33Hz).

Figure 13-5: Results of the modal analysis.

Run a separate Frequency study with five modes. Animate modes 2-5 to see that they correspond to local deformation of rubber legs.

Using the results of the modal analysis we know that there is only one mode in the range of operating frequencies 0-33Hz. Therefore, we reduce the number of modes to just one in **Frequency Options** and specify the range in **Harmonic Options** (Figure 13-6).

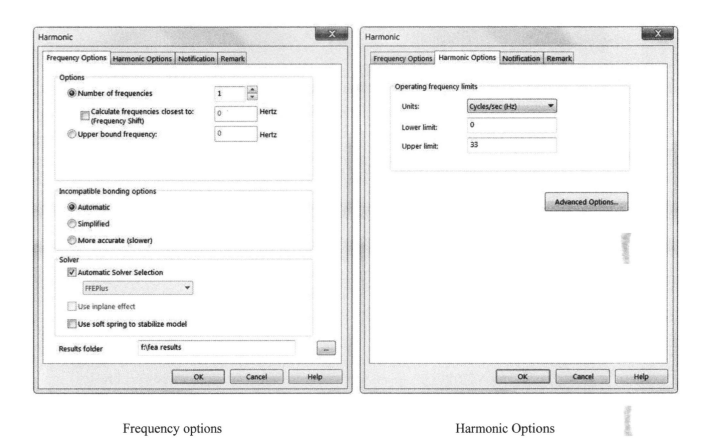

Frequency options   Harmonic Options

Figure 13-6: Options of the Harmonic study.

The operating frequency limits correspond to the range 0-2000RPM.

Define Modal Damping as 5%, which corresponds to high damping in the rubber material. The damping definition is shown in Figure 13-5.

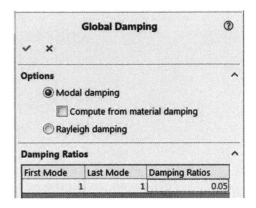

Figure 13-7: Definition of Modal Damping.

Damping only needs to be defined for one mode.

The load F has to be defined as a square function of angular velocity:

$$F = me\omega^2 .$$

Using a schematic representation of the centrifuge as in the assembly model CENTRIFUGE, we may represent the imbalanced force in a simplified manner by defining a load to the top face. Open the spreadsheet CENTRIFUGE, copy the highlighted table and follow the steps shown in Figure 13-8 to define the load.

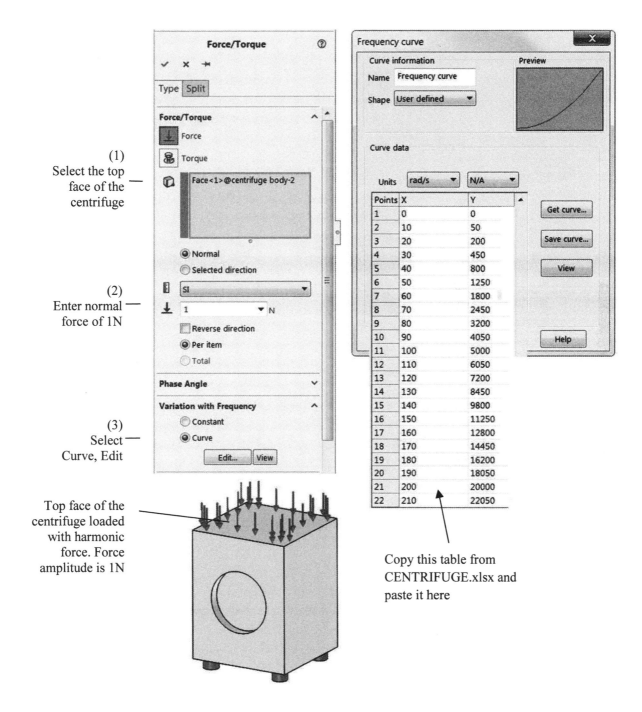

Figure 13-8: Definition of the imbalanced load.

1N is a multiplier to the Y column of the table copied from the CENTRIFUGE.xlsx spreadsheet. The Frequency curve window was modified to show all rows in the table.

Go to the CAD model to review the **Sensor** defined in one of the top corners of the centrifuge body. Define the **Results Options** shown in Figure 13-9.

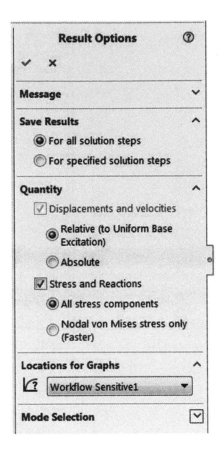

Figure 13-9: Definition of Results Options.

This is done in preparation to graphing the magnitude of displacement as a function of the angular velocity.

Review Sensor defined in SOLIDWORKS assembly model; it is located in one of the top corners but could have been placed anywhere on the centrifuge body.

Solve the **Harmonic** study and define a response graph showing the amplitude of displacements as a function of the angular velocity (Figure 13-10).

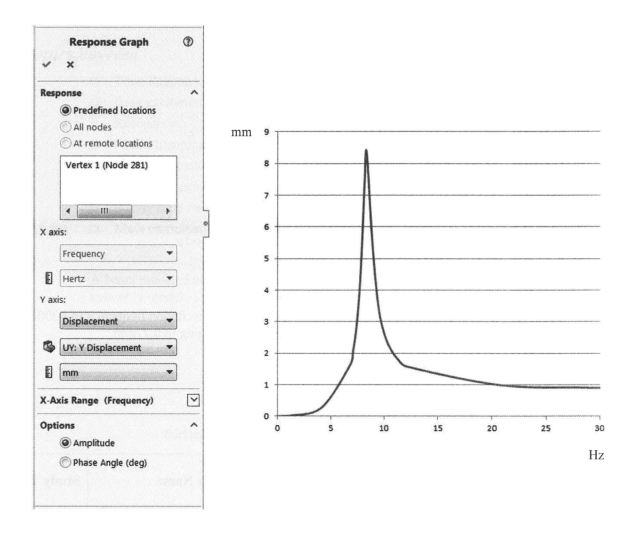

Figure 13-10: Amplitude of vertical displacement as a function of the excitation frequency.

The graph has been formatted in Excel.

The one peak in Figure 13-8 corresponds to the only natural frequency, 9.2Hz, present in the analyzed range of excitations. The maximum displacement amplitude is ~8mm and it happens when the excitation frequency equals the natural frequency 9.2Hz or 552RPM.

The model uses solid and surface geometry; surface geometry will be meshed with shell elements and solid geometry will be meshed with solid elements. To assure mesh compatibility, contact set must be defined between contacting entities of surface and solid geometries as shown in Figure 14-2.

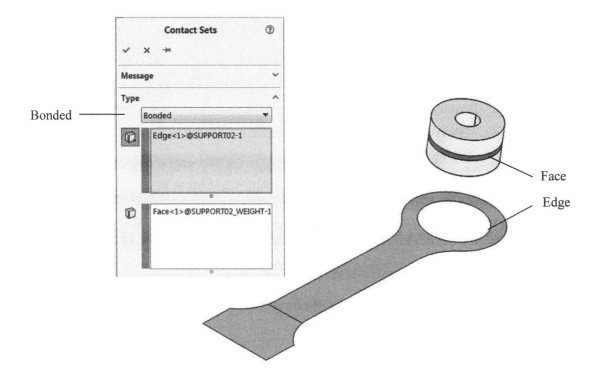

Figure 14-2: Bonded contact defined between the face of the solid and the edge of the surface. This applies to all studies completed in this chapter.

Model is shown in an exploded view.

The CANTILEVER BEAM is subjected to the harmonic base excitation with displacement amplitude of 0.05mm (Figure 14-3).

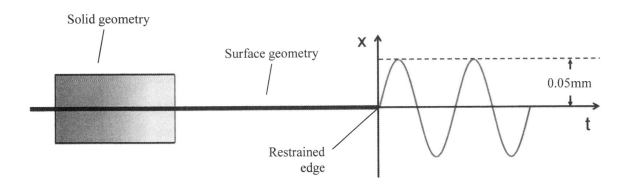

Figure 14-3: Harmonic base excitation applied to the CANTILEVER BEAM model.

The restrained edge performs harmonic oscillations in x direction with displacement amplitude of 0.05mm. The model is shown shorter to better fit this illustration.

We will investigate the beam response to the harmonic base excitation shown in Figure 14-3 when the frequency of excitation is close to the first natural frequency of the CANTILEVER BEAM model. This exercise will serve to study the differences between **Modal Time History** (**Time Response**) and **Harmonic** (**Frequency Response**) analyses. Both these analyses are based on the **Modal Superposition Method** and require the pre-requisite results of a **Frequency** (**Modal**) analysis.

Create a **Frequency** study titled *Modal* and define **Fixed** restraint as shown in Figure 14-1. Define contact set as shown in Figure 14-2. Define the shell element thickness as 3mm and mesh the model with the default element size.

Define **Result Options** as shown in Figure 14-8.

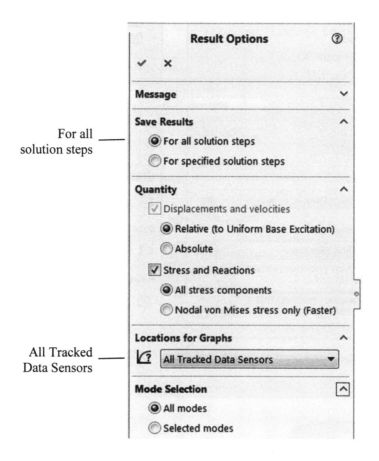

Figure 14-8: Result Options definition; make the above selections, accept default solution steps.

"All Tracked Data Sensors" selection means both sensors.

Run the solution and define the response graphs as shown in Figure 14-9.

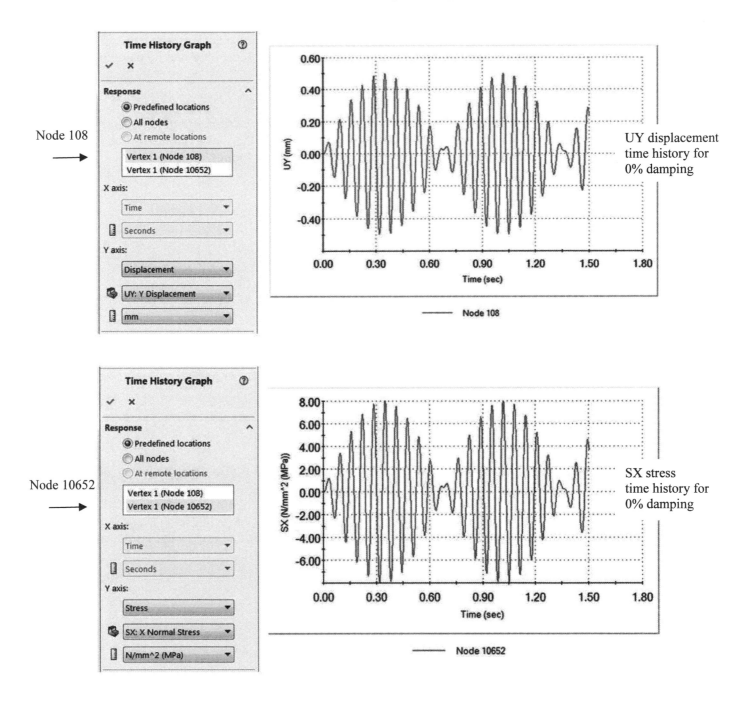

Node 108

Node 10652

Figure 14-9: Time History of UY displacement and SX stress in the locations of the sensors with 0% damping in the model.

Excitation frequency is 15Hz.

The total time of 1.5s corresponds to the time duration defined in the study Properties.

The stress sensor registers stress on the top of the shell element.

The graphs in Figure 14-9 illustrate the phenomenon called **Beating**. The response amplitudes periodically increase to the maximum magnitude and then decrease to zero. These changes happen for as long as the base excitation is active. The vibration response never reaches a steady state where the amplitude of vibration would become constant. Remember that no damping has been defined in this study.

To study the effect of damping, copy study *01 time response 0%* into study *02 time response 4%*. The only difference between these two studies will be the damping. In study *02 time response 4%* define **Modal Damping** as 4% (Figure 14-10).

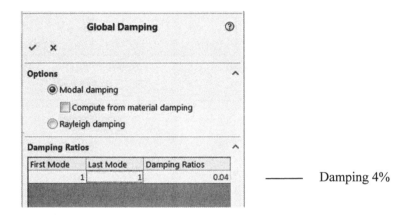

Figure 14-10: Modal Damping definition.

Study is based on one mode; therefore, damping is also defined only for one mode.

Obtain the solution of study *02 time response 4%* and review the displacement and stress time history as shown in Figure 14-11.

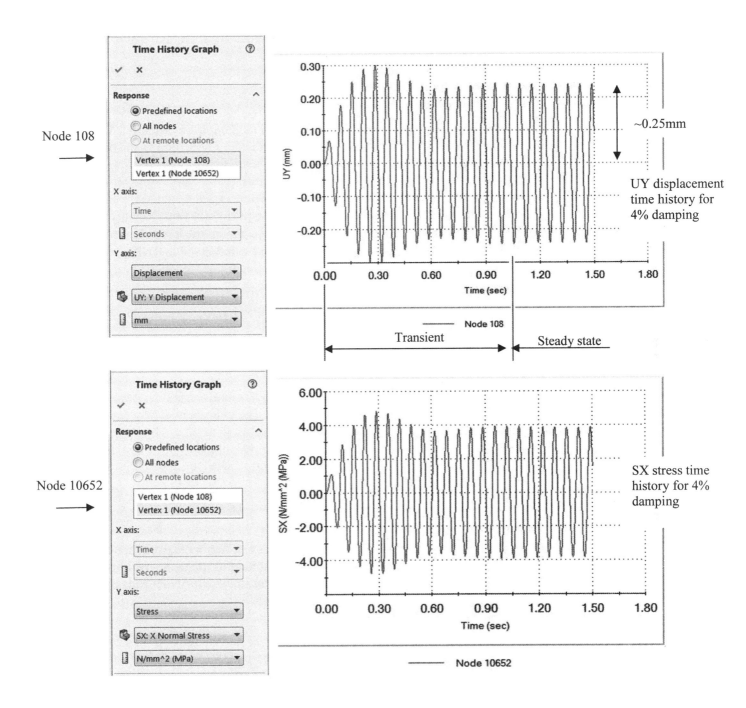

Node 108 →

Node 10652 →

Figure 14-11: Time History of UY displacement and SX stress in the locations of sensors with 4% damping; graph definition windows are the same as those shown in Figure 14-9.

Excitation frequency is 15Hz.

The transient response disappears after ~1s and the displacement amplitude reaches a steady state of ~0.25mm.

Here is a summary of both cases: without damping and with modal damping 4%.

Undamped response

The maximum UY displacement amplitude without damping is ~0.5mm. This maximum displacement amplitude repeats periodically for as long as the base excitation continues. The maximum displacement amplitude amplification is 0.5/0.05=10. The excitation frequency is very close to the natural frequency; the system is very close to resonance. The time distance between the maximum amplitudes is called the **Beating Period**. The **Beating Period** T depends on the natural frequency of the system f_n, and on the frequency of excitation f, which in our case gives the **Beating Period** equal to 0.67s:

$$T = \frac{1}{|fn - f|} = \frac{1}{|16.5 - 15|} = 0.67s$$

Damped response

The maximum UY displacement amplitude with 4% modal damping happens at t=0.28s when the amplitude reaches ~0.3mm. The maximum displacement amplitude amplification is ~6 and beating diminishes with time. The **Beating Period** remains the same as in the case of undamped vibration but it continues only for about 1s. Then the amplitude of vibration reaches the **Steady State** when the amplitude stabilizes at about 0.25mm.

Now, copy the study *02 time response 4%* into *03 time response 4% resonance*. In this new study change the frequency of harmonic excitation from 15Hz to 16.5Hz to make it equal to the first natural frequency within one decimal of accuracy. The displacement and stress time history in the locations of the sensors are shown in Figure 14-12.

Node 108

→

Node 10652

→

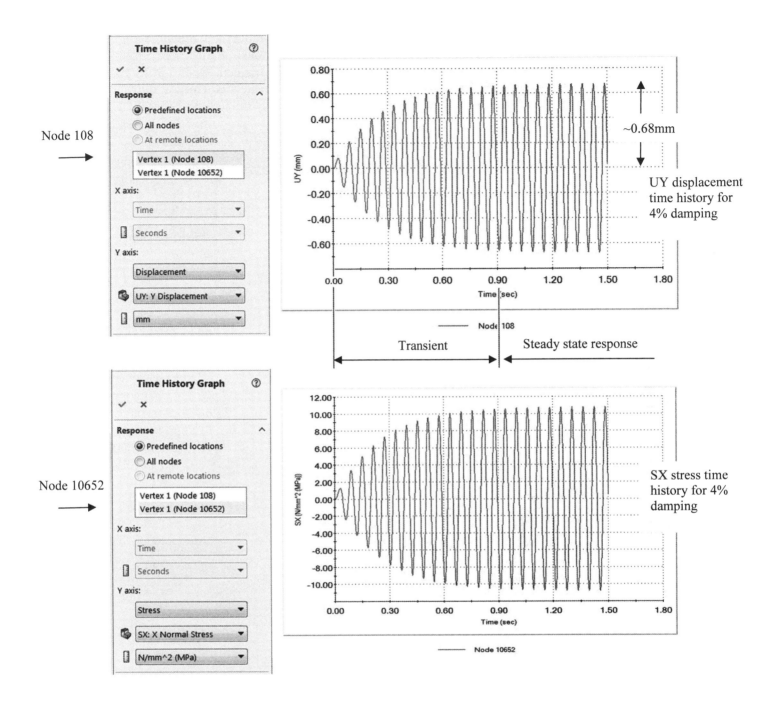

Figure 14-12: Displacement time history for harmonic excitation with a frequency of 16.5Hz which is equal to the natural frequency.

After ~ 0.9s the system reaches steady state, when the displacement amplitude stabilizes at about 0.68mm.

We'll now perform a **Frequency Response** study on the same model. To produce results that relate closely to the **Time Response** studies, we will again use only one mode, and a range of excitation frequencies of 0-30Hz to cover only the first natural frequency of 16.5Hz. Create a **Harmonic** study titled *04 frequency response 4%* and define the study properties as shown in Figure 14-13.

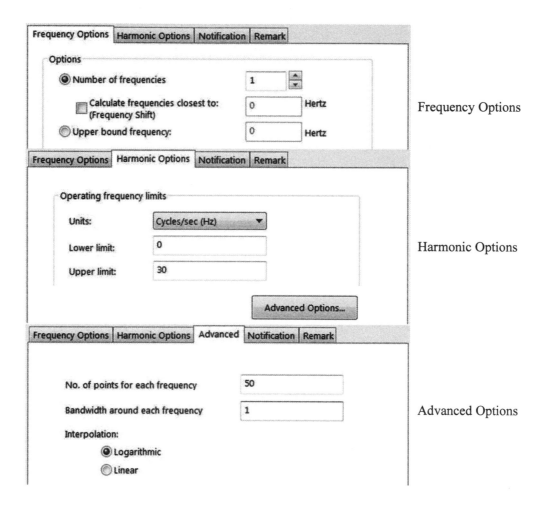

Frequency Options

Harmonic Options

Advanced Options

Figure 14-13: Harmonic study options.

Click Advanced Options in Harmonic Options to open the Advanced tab. Harmonic Options and Advanced Options windows have been modified in a graphics program to better fit this page.

In the **Frequency Options**, specify only one mode. This way the study will only be based on the first mode. In the **Harmonic Options**, specify the range of excitation frequencies as 0-30Hz. This captures both the natural frequency (16.5Hz) and the excitation frequency (15Hz) used in study *01 time response 0%* and in study *02 time response 4%*. In the **Advanced Options**, specify the number of points to be 50, a bandwidth of 1, and linear interpolation. This way the frequency range will be covered in 101 equal sized frequency steps with no bias around the natural frequency.

The number of performed frequency steps will be 101; 50 steps below the natural frequency, 50 steps above the natural frequency and one for the natural frequency; therefore, the total number of steps is 101.

Define **Base Excitation** and **Result Options** the same as in the **Time Response** studies. Define a **Modal Damping** of 4%. Run the study and construct the **Response Graph** showing the UY displacement amplitude as a function of the excitation frequency. Export the response graph to Excel and format it as shown in Figure 14-14.

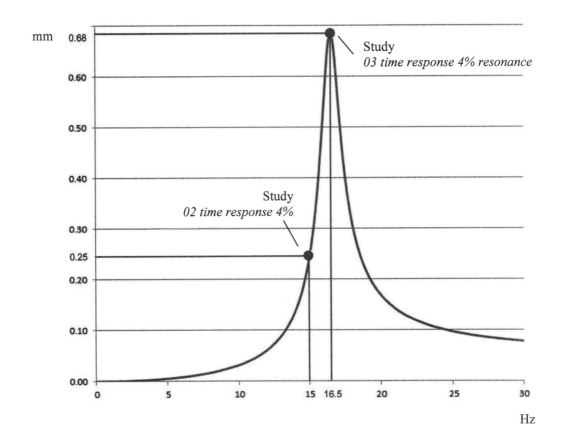

Figure 14-14: Displacement amplitude as a function of the excitation frequency. This is a displacement amplitude frequency response graph.

Two dots indicate the steady state solutions of two time response studies with 4% damping.

Figure 14-14 shows the steady state displacement amplitude for all frequencies within the range of 0-30Hz, including the frequency of 15Hz which was the excitation frequency in the **Time Response** study *02 time response 4%* and 16.5Hz, which was the excitation frequency in **Time Response** study *03 time response 4% resonance*.

Review the graph in Figure 14-14 and notice that each point on the **Frequency Response** curve corresponds to the steady state response that can be found using a **Time Response** study. The graph in Figure 14-14 could be constructed using steady state results of a large number (here one hundred and one) of time response analyses, clearly a very time consuming approach. The **Frequency Response** analysis provides these results within one study.

Summary of studies completed

Model	Configuration	Study Name	Study Type
CANTILEVER BEAM.sldasm	*Default*	*Modal*	Frequency
		01 time response 0%	Modal Time History
		02 time response 4%	Modal Time History
		03 time response 4% resonance	Modal Time History
		04 frequency response 4%	Harmonic

Figure 14-15: Names and types of studies completed in this chapter.

15: Vibration absorption

Topics covered

- ❏ Torsional vibration
- ❏ Resonance
- ❏ Modal damping
- ❏ Vibration absorption
- ❏ Frequency Response

Procedure

Open assembly model **VIB ABSORBER** and examine three configurations shown in Figure 15-1.

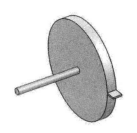 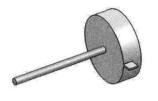 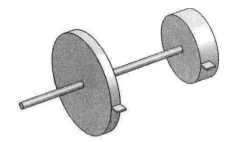

Configuration
01 one disk

Configuration
02 absorber

Configuration
03 full model

Figure 15-1: Three configurations of the VIB ABSORBER assembly.

Little tabs on both disks help to place sensors and aid in visualizing modes of vibration.

The assembly comes with sensors defined as shown in Figure 15-2.

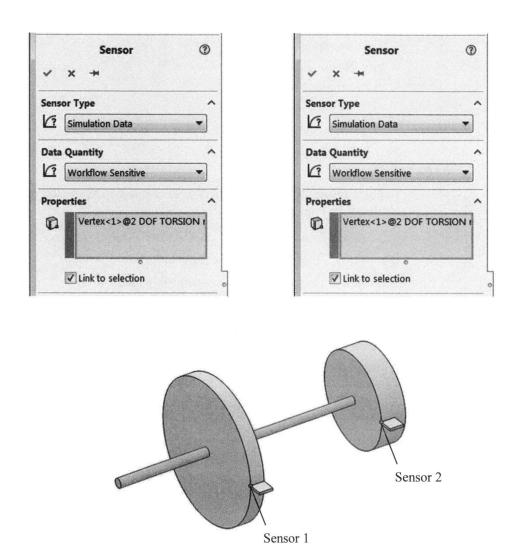

Figure 15-2: Sensor locations in the VIB ABSORBER model.

Sensors are located at the base of each tab.

We start an analysis in the *01 one disk* configuration to model the torsional vibration of one disk subjected to a harmonic torque load. The torque amplitude is 10Nm and does not change with the frequency of oscillations which is 0-80Hz. While 0-80Hz is the range of excitation frequencies, we are particularly interested in the steady state displacement amplitude at a frequency equal to 28Hz. For this purpose we use a **Harmonic** study.

Create a **Harmonic** study titled *01 one disk*. The disk is restrained as shown in Figure 15-3. The cylindrical faces of the shaft and disk are restrained in the radial direction using an **On Cylindrical Faces** restraint. The end face of the shaft has a **Fixed** restraint applied.

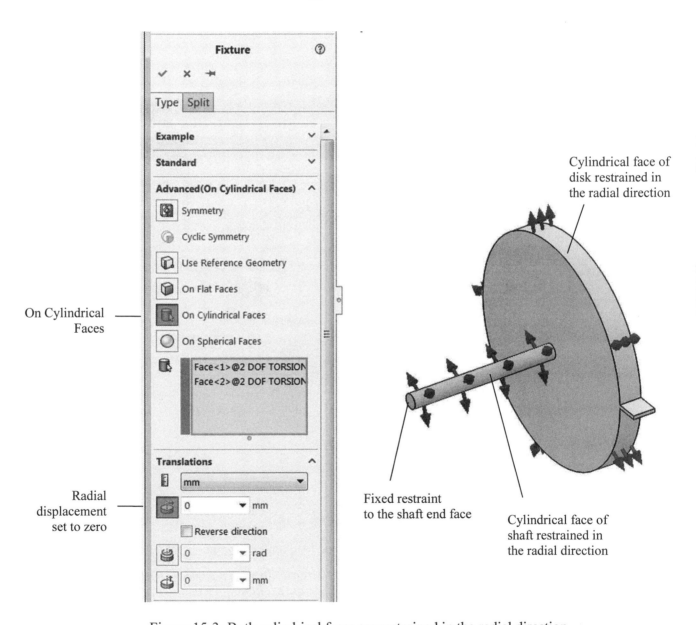

Figure 15-3: Both cylindrical faces are restrained in the radial direction.

The Fixed restraint definition window and symbol are not shown.

As a result of the restraints shown in Figure 15-3, the disk has only one mode of vibration in the range 0-80Hz. This is a torsional mode with a frequency of 28.1Hz. You may want to confirm this by running a modal analysis.

The disk is subjected to an oscillating torque of 10Nm as shown in Figure 15-4.

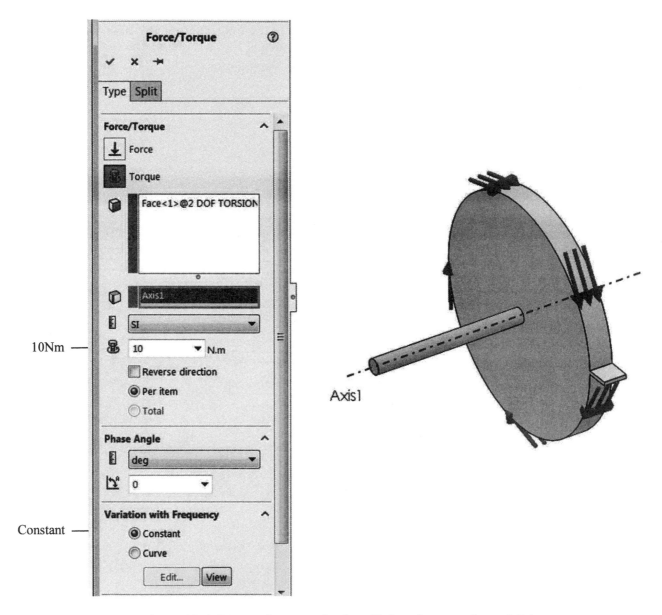

Figure 15-4: Harmonic torque load applied to the outer face of disk.

Selecting Constant for Variation with Frequency means that the torque amplitude does not change with the excitation frequency.

Define study properties as shown in Figure 15-5.

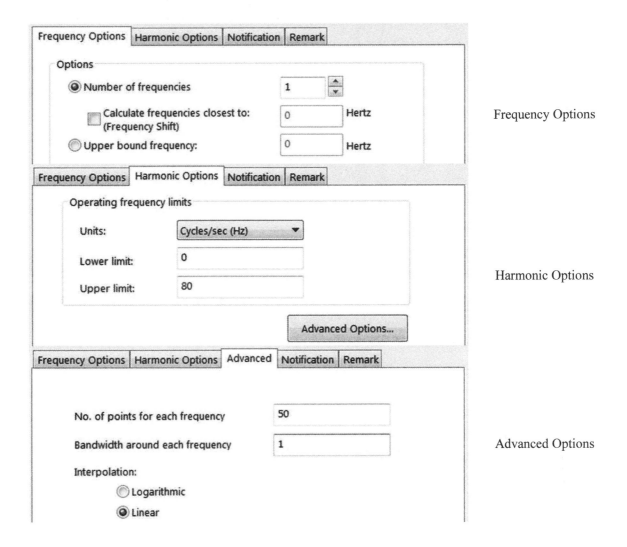

Figure 15-5: Properties of Harmonic study *01 one disk*.

One mode is selected in Frequency Options.

Frequency range 0-80Hz is specified in Harmonic Options.

50 points for each frequency, bandwidth 1 and linear interpolation are specified in Advanced Options.

Considering that only one mode is specified, 50 points defined in **Advanced Options** means that 50 frequency steps will be performed before reaching the natural frequency, another 50 after that and one for the natural frequency. Linear interpolations means that frequency steps won't be biased towards natural frequencies but will be evenly distributed over the frequency range 0-80Hz.

Define modal damping as 1% (Figure 15-6).

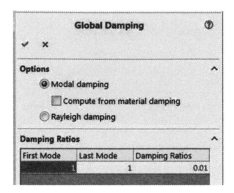

Figure 15-6: Modal damping defined as 1%.

Only one mode is shown in the Global Damping window because only one mode has been specified in the Frequency Options.

Define the **Result Options** as shown in Figure 15-7.

For all ——
solution steps

Sensor 1 ——

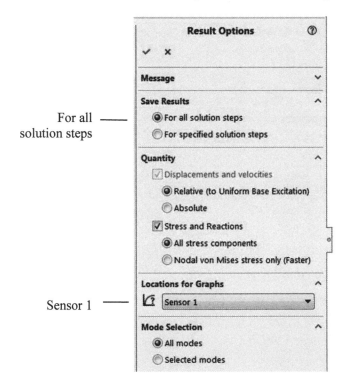

Figure 15-7: Definition of Result Options.

Only one sensor is available in the 01 one disk study; it is associated with configuration 01 one disk.

Mesh the model with the default element size and obtain the solution. Define a **Response Graph** as shown in Figure 15-8:

Vertex 1 is in
Sensor 1 location

UY
displacement
component

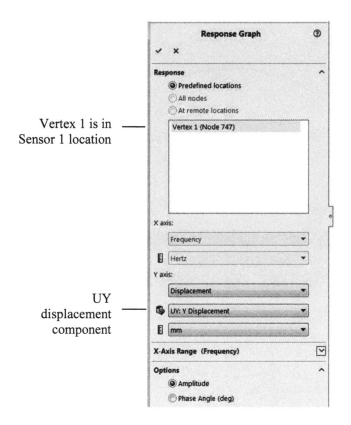

Figure 15-8: Definition of the displacement amplitude response graph.

We use a linear displacement on the circumference of the disk as a measure of angular rotation.

The UY displacement component selected in the **Response Graph** in Figure 15-8 is a linear displacement in the circumferential direction. If desired, it can be translated into angular displacement as shown in Figure 15-9.

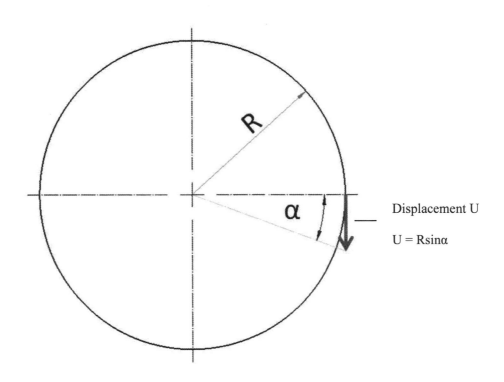

Displacement U

$U = R\sin\alpha$

Figure 15-9: UY displacement can be translated into an angular displacement.

The equation is valid only for small displacements.

The UY displacement amplitude frequency response graph is shown in Figure 15-10.

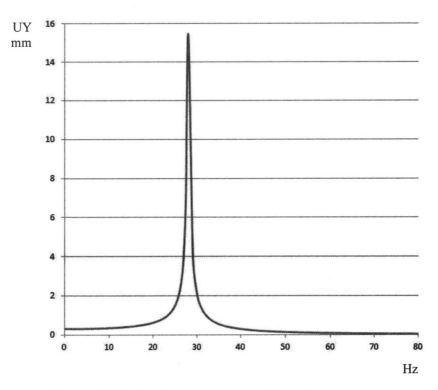

Figure 15-10: UY displacement amplitude frequency response. The peak corresponds to the natural frequency of 28.1Hz.

This graph has been formatted in Excel.

Damping in the system is low (1%); therefore, the maximum displacement amplitude happens when the excitation frequency equals the natural frequency. In other words, the system experiences resonance at 28.1Hz.

Now, let's imagine a situation where the disk is subjected to harmonic oscillations with a frequency of 28.1Hz which is the resonant frequency. As shown in Figure 15-10, the amplitude of linear displacements on the circumference is 15.5mm which translates into the angular rotation 5.9°.

We find this amplitude of vibration too large and want to reduce it. We'll do that by attaching a **vibration absorber**, the natural frequency of which will be equal to the frequency of excitation. Notice that this is the frequency of the absorber before it is attached to the disk.

We will first investigate the absorber before attaching it to the disk. Change to configuration *02 absorber*. Create a **Frequency** study *02 absorber*; apply restraints to the shaft and to the disk the same way as shown in Figure 15-3 and find the first mode of vibration to confirm that it is a torsional mode with a frequency of 28.1Hz (Figure 15-11).

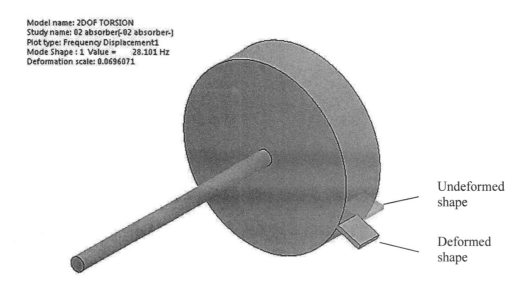

Model name: 2DOF TORSION
Study name: 02 absorber[-02 absorber-]
Plot type: Frequency Displacement1
Mode Shape : 1 Value = 28.101 Hz
Deformation scale: 0.0696071

Undeformed shape

Deformed shape

Figure 15-11: The first mode of vibration of the absorber: a torsional mode with a frequency of 28.1Hz.

The circular shape makes visualization of the torsional mode difficult. Tabs help the visualization.

The apparent growth of the disk diameter visible during modal animation is caused by the fact that displacement trajectories are straight lines tangent to the circumference of disk. This is an inherent property of any linear analysis.

Now we are ready for the main part of this exercise where we analyze the effect of the vibration absorber attached to the disk. Switch to assembly configuration *03 full model* and create a **Harmonic** study titled *03 full model*. In this configuration, the assembly consists of the disk and the absorber.

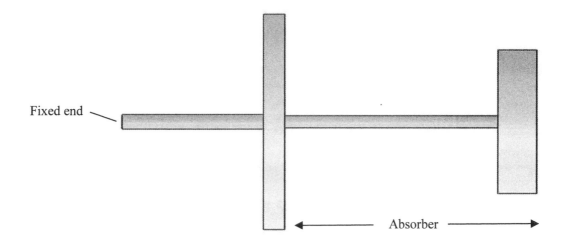

Fixed end

Absorber

Figure 15-12: The assembly model with the vibration absorber attached to the disk.

Disk is subjected to oscillating torque as shown in Figure 15-4.

Cylindrical faces of two disks and two shafts are restrained in the radial direction.

Define a **Harmonic** study with properties shown in Figure 15-13; two modes are now specified.

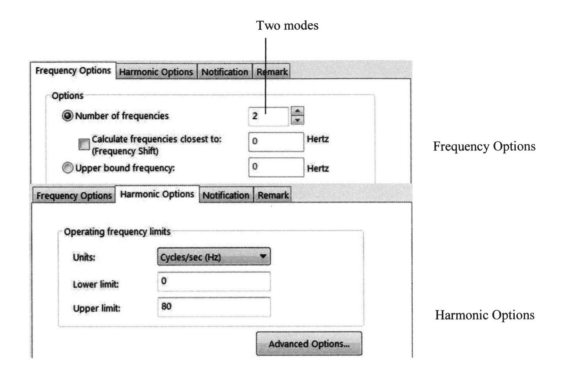

Figure 15-13: Properties of the Harmonic study for the model in the 03 full model configuration.

Advanced Options are not shown; make them the same as in Figure 15-5.

Apply restraints as explained in Figure 15-12. Define a harmonic torque excitation the same as in Figure 15-4 and **Results Options** as shown in Figure 15-14.

All Tracked
Data Sensors

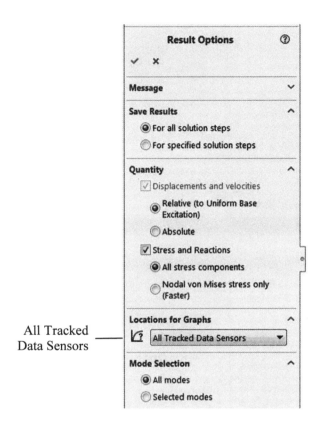

Figure 15-14: Results Option for the *03 full model* configuration.

"All Tracked Data Sensors" means Sensor1 and Sensor2.

The system has two natural frequencies in the range of 0-80Hz. Modal frequencies and shapes are shown in Figure 15-15.

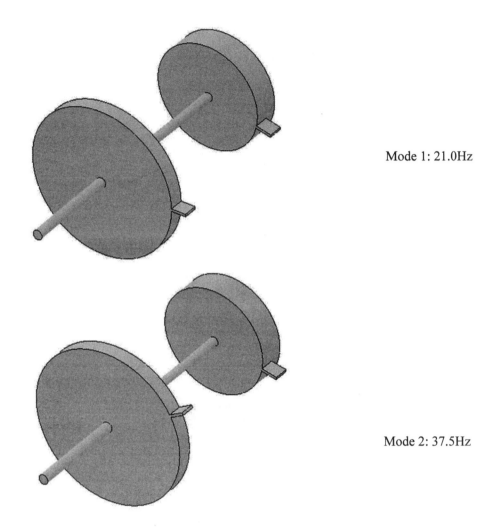

Mode 1: 21.0Hz

Mode 2: 37.5Hz

Figure 15-15: Two modes of vibration in the 0-80Hz range are torsional modes.

Undeformed shapes are not shown.

Animate modal shapes to see that in mode 1 both disks oscillate in the same direction and in mode 2 disks oscillate in the opposite directions.

Run the solution and notice that it proceeds in 151 steps because the **Advanced Options** in the study properties specify 50 points for each frequency (Figure 15-16).

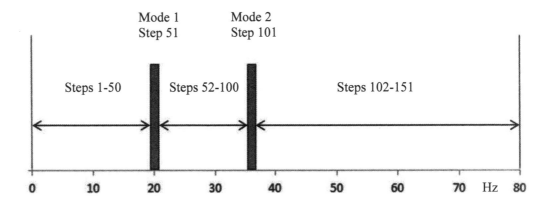

Figure 15-16: The range of frequencies is divided in three parts: 50 steps each.
The total step count is 151.

Once the solution completes, construct the **Response Graph** showing UY displacement amplitude as a function of the excitation frequency.

Response graphs formatted in Excel are shown in Figure 15-17.

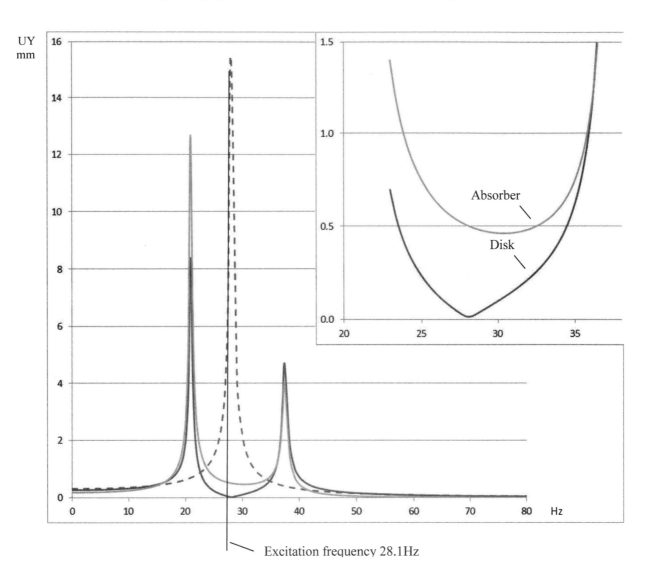

Figure 15-17: A summary of UY amplitude frequency response from the single disk study (dashed line) in *study 01 one disk*, and the disk with the absorber in study *03 full model*.

The main graph summarizes results from studies *01 one disk* (dotted line) and *03 full model* (solid lines). The dashed line repeats the results shown in Figure 15-10. The insert shows results from study *03 full model* in the narrow range of frequencies 24-36Hz.

It can be seen that attaching the absorber de-tunes the system from the resonant frequency of 28.1Hz. The system now has two resonant frequencies in the range 0-80Hz: 21.0Hz and 37.5Hz as shown by the two peaks in Figure 15-17 and modal shapes in Figure 15-14.

The results of the **Harmonic** study *03 full model* provide an overview of the system response in the specified range of frequencies. We are particularly interested in the frequency of 28.1Hz which is the frequency of the torque excitation and the frequency of the absorber before attaching it to the main disk. As Figure 15-17 illustrates it, when the absorber is attached to the main disk and the main disk is excited with a 28.1Hz frequency, the vibration of the main disk stops at this frequency of excitation; see the insert in Figure 15-17 showing very low amplitude of vibration. We say that the vibration has been "absorbed" by the added vibrating system, and hence the name **Vibration Absorber**.

Controlling vibrations by means of adding an absorber is practical only in cases where the frequency of excitation is constant and adding components is possible. Assembly model VIB ABSORBER presents an idealized system performing angular vibration. We may treat it as a discrete vibrating system because:

1. The mass moment of inertia of shafts is low as compared to the mass moment of inertia of disks (the main disk and absorber). The torsional stiffness of disks is high as compared to the torsional stiffness of shafts. Therefore, we may assume that shafts are responsible for stiffness properties and disks are responsible for inertial properties of this system.

2. The system is subjected to restraints in such a way as to have two separate torsional modes of vibration within the range of the investigated frequencies.

The analysis conducted on VIB ABSORBER assembly model illustrates the principle of vibration absorbing which applies to systems performing linear and angular vibration.

The analysis does not address important problem such as stress or fatigue. You may want to analyze stresses as function of frequency to see that yield strength of material is exceeded at resonance.

Summary of studies completed

Model	Configuration	Study Name	Study Type
VIB ABSORBER.sldasm	*01 one disk*	*01 one disk*	Harmonic
	02 absorber	*02 absorber*	Frequency
	03 full model	*03 full model*	Harmonic

Figure 15-18: Names and types of studies completed in this chapter.

16: Random vibration

Topics covered

- ❑ Random vibration
- ❑ Power Spectral Density
- ❑ RMS results
- ❑ PSD results
- ❑ Modal excitation

Random vibration

Random vibrations are non-periodic. Knowing the history of random vibration, we can predict the probability of occurrence of acceleration, velocity and displacement magnitudes, but cannot predict the precise magnitude at a specific instance in time.

Random vibration is composed of a continuous spectrum of frequencies. The huge amount of time history data makes it impractical or impossible to solve random vibration problems using time response analysis.

For most structural vibrations, the excitation, such as a base acceleration, alternates about zero. Consequently, mean values characterizing the excitation, as well as responses to that excitation such as displacement or stress, are equal to zero and can't be used to characterize random vibration responses. For this reason, results of a random vibration analysis are given in the form of Root Mean Square (RMS) values.

To explain the concept of an RMS value, refer to the graph in Figure 16-1 which shows the acceleration time history (acceleration as a function of time) of random vibration expressed in units of gravitational acceleration [G].

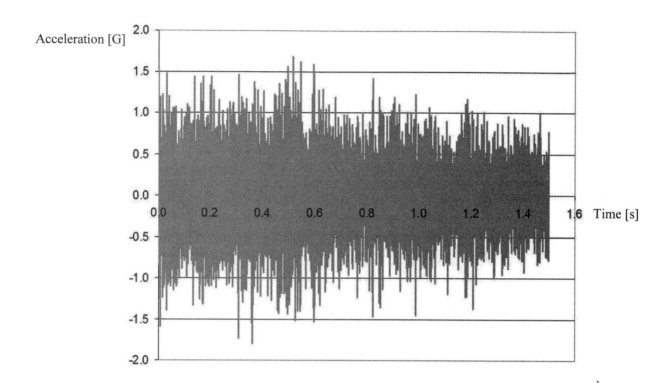

Figure 16-1: An example of acceleration time history data collected during 1.5s.

Considering the sampling rate of 5000 samples per second, this time history curve contains 7500 data samples. This acceleration time history has a zero mean.

The acceleration time history shown in Figure 16-1 has a zero mean value. However, if we multiply the function by itself, we obtain a function with a positive value. This squared function will be well suited to characterize the acceleration time history because its mean value will not be zero. This mean value of square acceleration time history is the mean square value and has units of $[G^2]$.

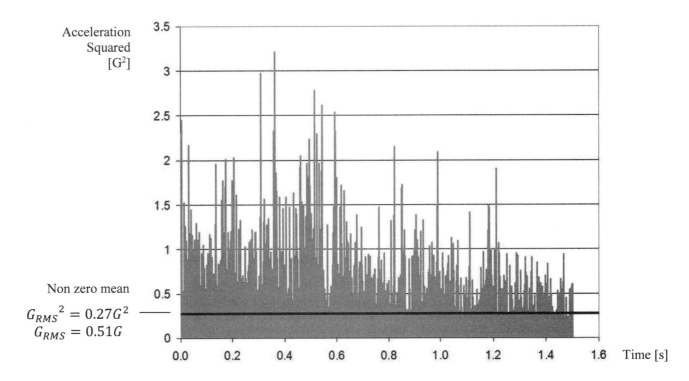

Acceleration Squared $[G^2]$

Non zero mean

$$G_{RMS}^2 = 0.27G^2$$
$$G_{RMS} = 0.51G$$

Time [s]

Figure 16-2: Squaring the acceleration time history function shown in Figure 16-1 produces a function with a non-zero mean.

Non-zero mean can be used to characterize the acceleration time history.

As shown in Figure 16-2, calculating the mean square value gives $G_{RMS}^2 = 0.27G$. The square root of the mean square value gives $G_{RMS} = 0.51G$.

The square root of the mean value is the root-mean-square (RMS) acceleration and has units of [G]. The same applies to RMS displacement, velocity, stress etc.

In random vibration, the magnitudes of acceleration, velocity, displacement etc. all follow a normal distribution. The RMS value corresponds to one standard deviation σ characterizing the normal distribution. To explain this, we refer again to Figure 16-2. The acceleration, as characterized by the given acceleration time history, has a 68% probability of remaining between -0.51G and +0.51G. Consequently, it has a 32% probability of being less than -0.51G or more than 0.51G.

Acceleration Power Spectral Density

Let's assume that the acceleration time history in Figure 16-1 is a stationary random process where probability numbers characterizing this process do not change with time. In this case, the acceleration time history can be used to calculate the **Acceleration Power Spectral Density (PSD)** curve: the variation of any property with respect to frequency is called "spectrum."

The overall G_{RMS}^2 of random vibrations shown in Figure 16-2 is $0.27G^2$. However, random vibration is composed of a large number of frequencies. Let us say we wish to investigate G_{RMS}^2 individually for a number of frequencies in the range from 0 to 2000Hz. Therefore, we divide the 0-2000Hz range into 20 sections (bins), each 100Hz wide and calculate G_{RMS}^2 characterizing each section by filtering out all frequencies falling outside of the section (Figure 16-3).

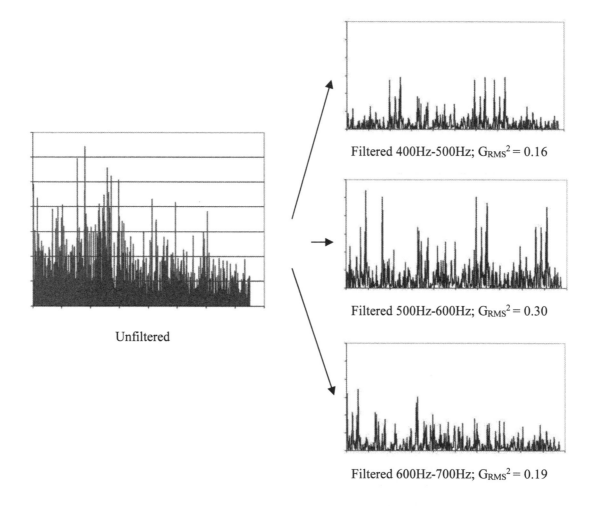

Unfiltered

Filtered 400Hz-500Hz; $G_{RMS}^2 = 0.16$

Filtered 500Hz-600Hz; $G_{RMS}^2 = 0.30$

Filtered 600Hz-700Hz; $G_{RMS}^2 = 0.19$

Figure 16-3: G_{RMS}^2 calculated individually for specified frequency ranges.

The graph on the left shows the squared acceleration time history from Figure 16-2. Only three frequency ranges (sections) are illustrated here for brevity.

Having found G_{RMS}^2 values obtained for each frequency range, we can now calculate individual "densities" of G_{RMS}^2 in each section by dividing G_{RMS}^2 in each section by the width of the section. Results obtained for all sections may be plotted as a function of the frequency in the center of each section. This function is called the **Acceleration Power Spectral Density** (Figure 16-4).

Band pass filter	Band center	G_{RMS}^2	Bandwidth	Acceleration PSD
	Hz	G^2	Hz	G^2/Hz
400Hz - 500Hz	450	0.16	100	0.0016
500Hz - 600Hz	550	0.30	100	0.0030
600Hz - 700Hz	650	0.19	100	0.0019

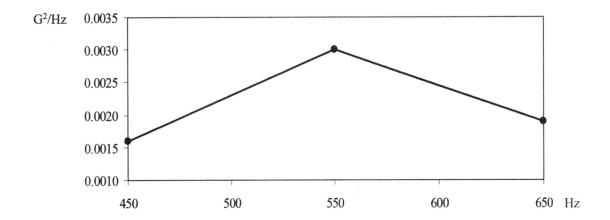

Figure 16-4: Constructing the Acceleration Power Spectral Density function (PSD).

Three points of the Acceleration PSD curve have been calculated by dividing G_{RMS}^2 in each section (each frequency range) by the width of the section.

The Acceleration Power Spectral Density (PSD) allows for a compression of data and is commonly used to characterize a random process. In particular, mechanical vibrations are commonly described by the Acceleration Power Spectral Density, which is easily generated by testing equipment. Design specifications and test results of devices subjected to random vibration are typically given in the form Acceleration PSD.

Analysis of random vibration of a hard drive head

Having completed this short introduction of random vibration, we now begin a random vibration analysis of a hard drive head. Open the assembly HD HEAD and examine two configurations: *01aligned* and *02 misaligned*.

The assembly contains only one part: HD HEAD; why can't we analyze the part instead of the assembly? In a **Random** study, the base excitation may only be applied along only one of the global directions X, Y, Z. To apply a base excitation in any other direction, we must place the analyzed part in the desired position respective to the global coordinate system. The assembly enables positioning of the part with respect to the global coordinate system. In configuration *01 aligned* the HD HEAD is aligned with the global coordinate system. In configurations *02 misaligned*, it is rotated 45° about the X axis.

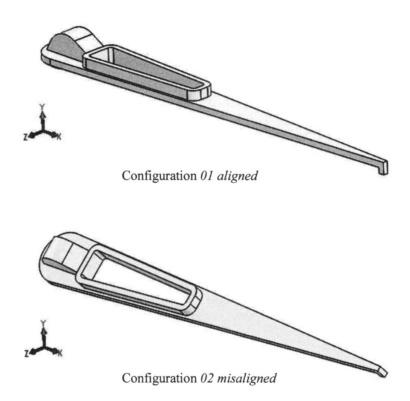

Configuration *01 aligned*

Configuration *02 misaligned*

Figure 16-5: Two assembly configurations are different only in the part orientation relative to the reference system.

In configuration 01 aligned, all part reference planes are aligned with the corresponding assembly reference planes. In configuration 02 misaligned, the part is rotated 45° about the global X axis.

We'll start in configuration *01 aligned*. Create a **Frequency** study with properties shown in Figure 16-6; call the study *Modal*.

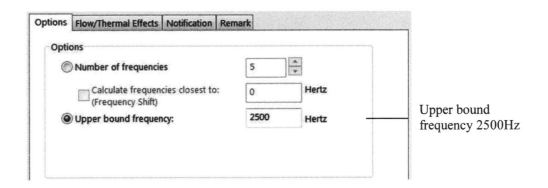

Upper bound frequency 2500Hz

Figure 16-6: Properties of the Modal study.

All frequencies in the range of 0-2500Hz will be calculated.

Apply restraints as shown in Figure 16-7.

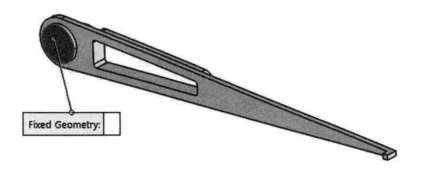

Fixed Geometry:

Figure 16-7: Fixed restraint applied to the hard drive head model.

Define **Mesh Control** as shown in Figure 16-8 and mesh the model with the default element size.

Controlled faces —

Element size on the controlled — faces: 0.2mm

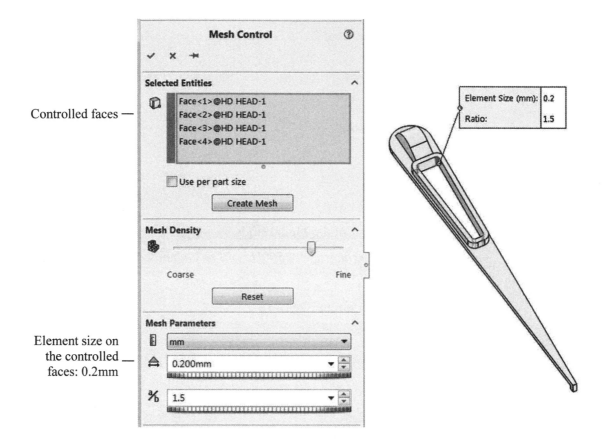

Figure 16-8: Mesh controls are applied to four round fillets.

Mesh controls specifies elements the size of 1/6 of the element size used to mesh the rest of the model.

Having defined mesh controls, use the default global element to mesh the model. If displacement results were our only objective, the default mesh would be acceptable for both **Frequency** and **Random** studies. However, since in the next study we intend to analyze displacements and stresses, mesh controls are required to ensure correct element shape and size in the area of stress concentrations. Prior to this exercise, a static analysis was run to find where mesh controls should be specified.

Solve the *Modal* study and review the results shown in Figure 16-9. Notice that there are four modes of vibration within the specified range of frequencies.

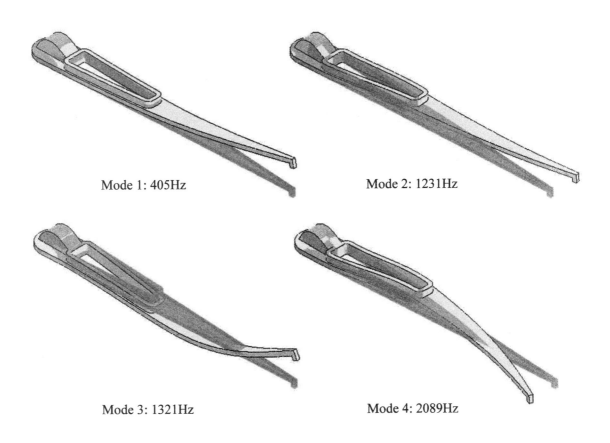

Mode 1: 405Hz   Mode 2: 1231Hz

Mode 3: 1321Hz   Mode 4: 2089Hz

<u>Figure 16-9: Modes of vibration within the range of 0 – 2500Hz.</u>

Vibration in mode 1 and 3 take place in the XY plane; vibrations in mode 2 and 4 take place in the XZ plane. The undeformed model is overlaid on the modal shape plots.

Proceeding to the analysis of the **Random** study, we could create a new **Dynamic** study independent from the completed **Frequency** study, with the **Random** option selected. However, a **Frequency** analysis would then have to be repeated within a **Dynamic Random** study. To avoid this repetition, we can copy the results of the **Frequency** study into a **Dynamic Random** study as shown in Figure 16-10.

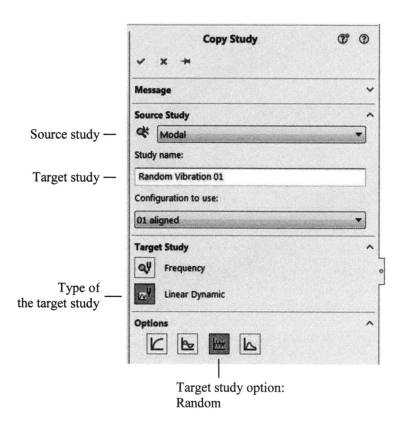

Figure 16-10: Results of a Frequency study can be copied to a new Dynamic study with Random option.

Name the study Random Vibration 01. Copying the Frequency study into a Dynamic Random study also copies mesh and restraints information.

The required properties of the **Random Vibration** study are shown in Figure 16-11.

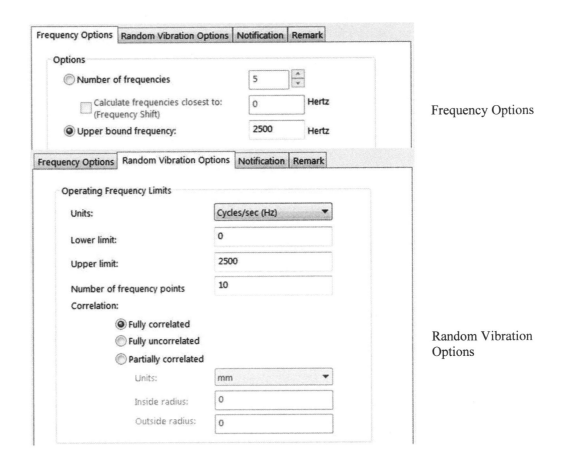

Frequency Options

Random Vibration Options

Figure 16-11: Properties of the Random Vibration study.

The Frequency Options specify all modes in the range of 0-2500Hz to be considered in the analysis of random vibrations. The Frequency Options have been copied from Modal study; you don't have to define them.

In the Random Vibration options, specify the Upper limit as 2500Hz to investigate responses to Random Vibration in the frequency range from 0 to 2500Hz.

Define **Global Damping** as shown in Figure 16-12.

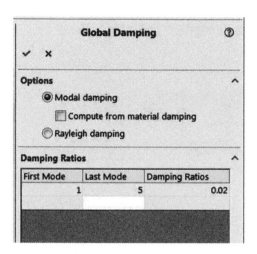

Figure 16-12: Global damping definition.

Global damping 2% is defined for all modes used in the analysis.

Assigning the same damping ratio to all modes represents a simplified and conservative approach. In most cases damping for higher modes will be higher than for lower modes.

The load on the hard drive head comes from random excitation of the base in the global Y direction. Follow Figure 16-13 to define the Acceleration PSD. The Acceleration PSD has been obtained from testing.

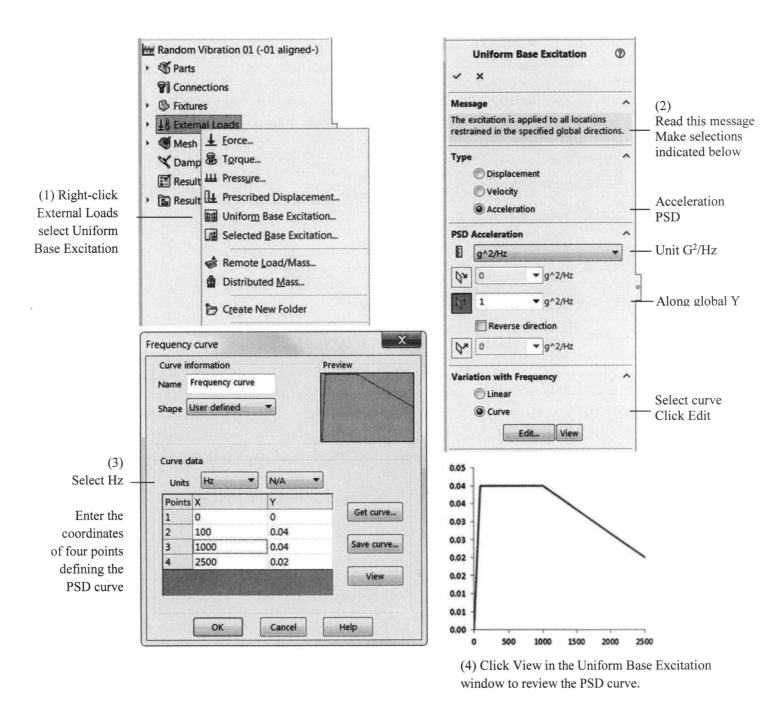

(1) Right-click External Loads select Uniform Base Excitation

(2) Read this message Make selections indicated below

Acceleration PSD

Unit G²/Hz

Along global Y

Select curve Click Edit

(3) Select Hz

Enter the coordinates of four points defining the PSD curve

(4) Click View in the Uniform Base Excitation window to review the PSD curve.

Figure 16-13: Uniform Base Excitation defined as the Acceleration PSD in the global Y direction acting on all restraints present in the model. In our model only one restraint is present.

Follow the above steps to create the Acceleration PSD curve. The graph has been formatted in Excel.

Define two **Sensors** as shown in Figure 16-14.

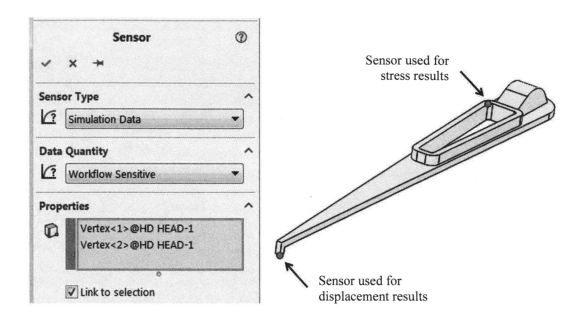

Figure 16-14: Sensor locations; two vertices are selected: at the tip of the head and at the end of the fillet.

HD HEAD assembly model comes with sensors already defined.

Define **Result Options** as shown in Figure 16-15.

All tracked Data Sensors ———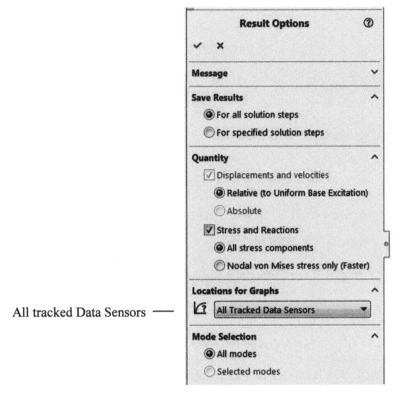

Figure 16-15: Result Options.

All tracked Data Sensors includes the two locations selected in the Sensor definition window in Figure 16-14.

Obtain the solution and analyze the RMS displacement results and the PSD displacement results.

Define UY displacement plot of RMS and PSD displacements as shown in Figure 16-16.

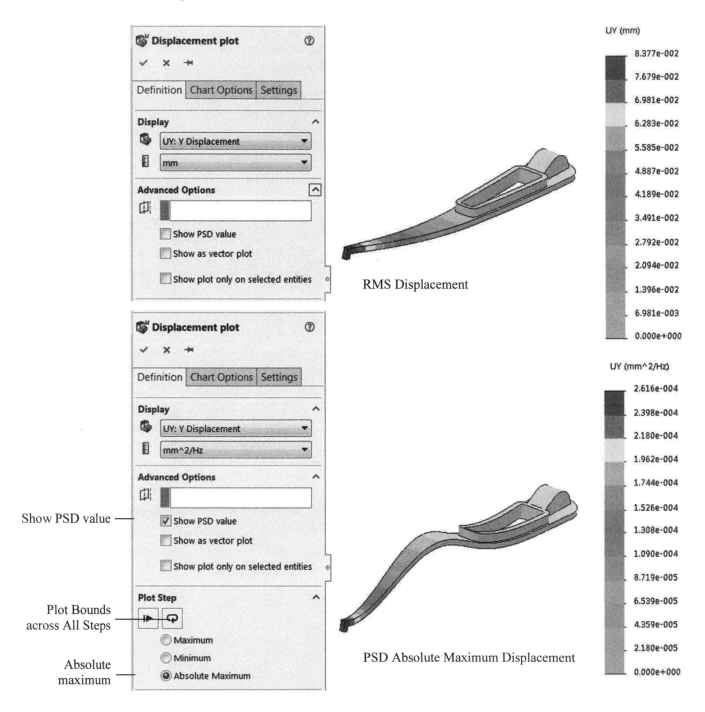

Figure 16-16: RMS displacement results and PSD Absolute Maximum UY displacement component.

The maximum RMS displacement is 0.084mm. The maximum PSD displacement is $0.00026mm^2/Hz$.

PSD displacements are displayed in units of mm^2/Hz for the **Absolute Maximum** found in the frequency range. The **Absolute Maximum** of PSD displacement corresponds to the first mode frequency 405Hz.

It is important to understand the meaning of results in a **Random Vibration** analysis. The displacement results in Figure 16-16 (top) are the RMS displacements. The maximum RMS displacement is 0.084mm meaning that the magnitude of displacement has a 68% probability of remaining under 0.084mm. The probability of the maximum displacement magnitude exceeding 0.084mm is 100% - 68% = 32%.

Remembering that the probability of a given displacement is defined by a normal distribution for which σ = 0.084mm, we can calculate the probability of displacement magnitude exceeding any defined value (Figure 16-17).

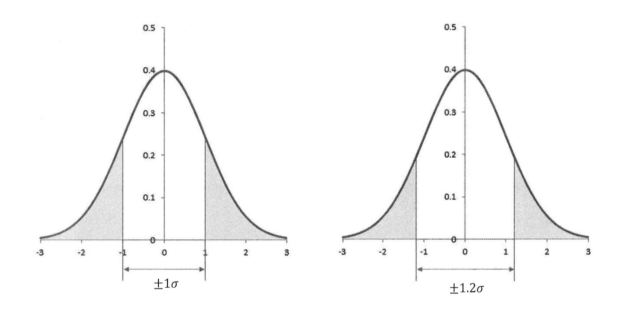

Figure 16-17: The total area under the normalized Gauss curve is 1. The probability of displacement magnitude exceeding ±σ (left) and ±1.2σ (right) is equal to the corresponding shaded areas.

The probability of the displacement magnitude exceeding 1 x RMS displacement (here 0.084mm) is given by the area outside ± 1σ which is 32% (left).

The probability of displacement magnitude exceeding 1.2 x RMS displacement (0.10mm) is given by the area outside ± 1.2σ which is 23% (right).

The same applies to all results of a Random Vibration analysis. The RMS von Mises stress result is shown in Figure 16-18.

Stress concentration

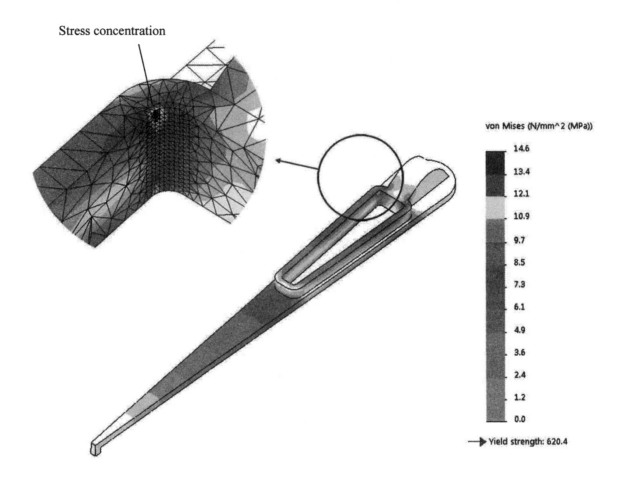

Figure 16-18: RMS von Mises stress plot.

The maximum von Mises stress has a 68% probability of remaining below 14.6MPa. Section Clipping is used in this plot; stress concentrations are present on both sides of the model.

Notice that stress singularities present in the model (sharp re-entrant edges) do not show as stress concentrations because of a large element size used for meshing. Refer to Figure 16-18 and compare the size of elements to the size of the stress concentration. Even with mesh controls applied, the mesh is at best marginal to model stress concentrations. Repeat the analysis using a more refined mesh.

The results of **Dynamic Random** analysis presented as RMS values provide one result for the entire frequency range of excitation. Results of Random Vibration analysis such as displacements or stresses may also be presented as PSD values, which are different for each excitation frequency. Examine the PSD options of displacements and stress result plots and notice that displacement results are given in mm^2/Hz, and stress results are given in MPa^2/Hz. These units are a consequence of the base excitation being defined as acceleration squared per frequency range, in our case G_{RMS}^2/Hz.

The most informative way to review PSD results is to graph them over the frequency range. Create two graphs showing PSD displacement of the tip of the head defined by the sensor shown in Figure 16-14. These two graphs are shown in Figure 16-19.

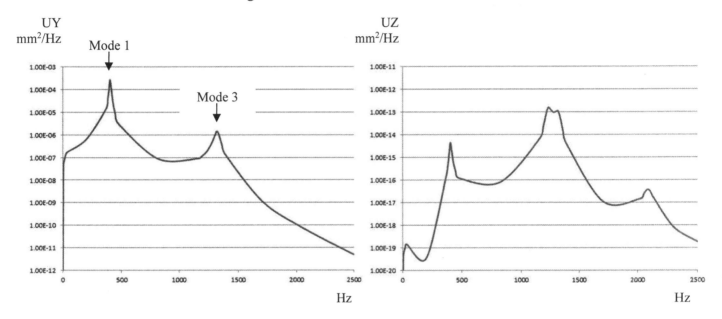

Figure 16-19: PSD UY and UZ displacement components of the tip as a function of excitation frequency in configuration *01 aligned.*

Notice a very different scale of Y axis in UY and UZ response graphs. These graphs have been formatted in Excel.

Modal shapes of mode 1 and mode 3 are aligned with the direction of base excitation. Indeed, upon examination of the UY response graph in Figure 16-19 we see that mode 1 and mode 3 are excited, even though the effect of mode 3 is visible only because a logarithmic scale is used. The UZ response graph does not show any modes excited. The peaks visible in the UZ graph are results of discretization error.

The area under the PSD UY response graph equals the RMS^2 displacement of the vertex where the sensor has been defined. This can be proven by integrating the PSD UY displacement function. If you wish to perform the numerical integration, then for better accuracy copy the study *Random Vibration 01* into *Random Vibration 01 small steps* and use settings shown in Figure 16-20.

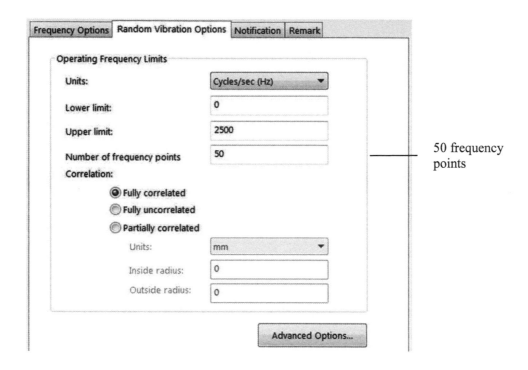

Figure 16-20: Settings of *Random Vibration 01 small steps* study.

Specify 50 frequency points.

Figure 16-21 shows the UY PSD response graph based on 251 frequency steps; it uses a linear scale on the Y axis. The results of numerical integration are shown in spreadsheet HD HEAD.xlsx.

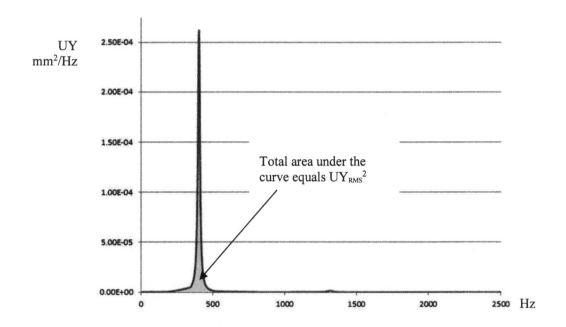

Figure 16-21: Correspondence between RMS and PSD UY displacement: Total area under the curve equals UY_{RMS}^2.

See the spreadsheet HD HEAD.xlsx for details.

Now, change to configuration *02 misaligned*, create a **Random** study titled *Random Vibration 02* and repeat the analysis. Everything except the model position will be the same as in the study *Random Vibration 01*. Obtain the solution and construct PSD response graphs for UY and UZ displacement components of the tip. These graphs are shown in Figure 16-22 and are directly comparable to the graphs presented in Figure 16-19.

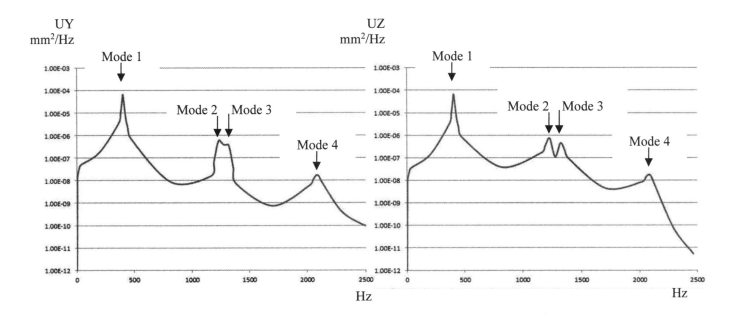

Figure 16-22: PSD UY and UZ displacement components of the tip as a function of excitation frequency in configuration *02 misaligned*.

Four peaks in each graph correspond to the four modes excited by the base excitation.

Upon examination of the graphs in Figure 16-22, we find that in configuration *02 misaligned*, all four modes are excited; this is easiest to show using a logarithmic scale. Mode 1 still dominates the vibration response.

Summary of studies completed

Model	Configuration	Study Name	Study Type
HD HEAD.sldasm	*01 aligned*	*Modal*	Frequency
		Random Vibration 01	Random Vibration
		Random Vibration 01 small steps	Random Vibration
	02 misaligned	*Random Vibration 02*	Random Vibration

Figure 16-23: Names and types of studies completed in this chapter.

17: Response spectrum analysis

Topics covered

- Non stationary random base excitation
- Seismic response analysis
- Seismic records
- Response spectrum method
- Generating response spectra
- Methods of mode combination

Non stationary excitation

Many random events are not stationary, meaning that parameters such as mean or variance do not remain the same but change with time. Important examples of a non-stationary process are earthquake and pyrotechnic shock. Figure 17-1 shows an acceleration time history of an earthquake recorded in California zone 4. The event duration is 31s; considering a sampling rate of 0.005s the time history consists of 6145 data points.

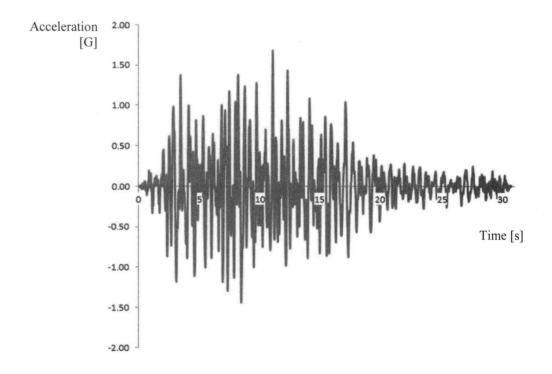

Figure 17-1: An example of acceleration time history of an earthquake.

Source: vibrationdata.com.

The earthquake is a non-stationary process and the methods of Random Vibration analysis cannot be used. At the same time, given the amount of data points, a **Time Response** analysis is impractical. For these reasons, a special analysis method called the **Response Spectrum Method** has been developed to analyze long duration non-stationary processes like the earthquakes or pyrotechnic shock events.

We will explain the concept of the **Response Spectrum Method** in the following steps:

1. A system vibrating in resonance can be described as a single degree of freedom harmonic oscillator system characterized by mass, stiffness and damping.

2. A response of a system with more than one resonant frequency can be represented as a combination of responses of harmonic oscillators, with each harmonic oscillator corresponding to a particular resonant frequency. Notice that this is the basis of the modal superposition method.

3. If the excitation frequency is equal to one of the structure's resonance frequencies, then the system response to that excitation is controlled only by system damping. Mass does not matter, stiffness does not matter; they have no impact on the system's response. The **only** thing controlling the system response is its damping!

Imagine that two vastly different systems, a harmonic oscillator and a bridge, have two important things in common: the resonant frequency and damping. Now imagine that both the harmonic oscillator and the bridge are excited by the same excitation that happens to have the same frequency as the resonance frequency of both the harmonic oscillator and the bridge. The vibration response (e.g. the maximum displacement) will be the same for the harmonic oscillator and the bridge!

4. Let's assume that the vibration response can be adequately modeled with the modal superposition method based, for example, on five modes. Using the observations made in point **3**, we may simplify the analysis very significantly. Rather than studying the response of the actual structure, we can study the response of five harmonic oscillators with natural frequencies corresponding to those five modes and associated damping being the same as the modes and damping of the structure we wish to analyze. In particular, if we study the structure's response to seismic excitation, then rather than testing the actual structure, we can just subject those five oscillators to the earthquake acceleration time history and record the maximum displacement, velocity, and acceleration of each oscillator. Next, we plot the maximum displacement, velocity, and acceleration of each oscillator as a function of the oscillator's frequency. This way a **Response Spectrum** curve is built. The maximum displacement recorded as a function of frequency is the **Displacement Response Spectrum**, the maximum velocity recorded as a function of frequency is the **Velocity Response Spectrum** and the maximum acceleration recorded as a function of frequency is the **Acceleration Response Spectrum**. The above reasoning may be extended to any number of harmonic oscillators.

5. The examined structure does not have to have the same resonant frequencies as our set of harmonic oscillators described in point **4**. If the resonant frequencies of the analyzed structure fall "in-between" frequencies of the oscillators used to construct the **Response Spectrum** curve, the structure response can be interpolated. So if resonant frequencies of the analyzed structures are known and we also know the **Acceleration Response Spectrum** curve that has just been constructed by examining the response of harmonic oscillators, then we can find out by interpolation, the maximum acceleration of the structure corresponding to each mode. Double integration will give the maximum displacement.

6. The information of interactions between modes has been lost in the above process. Therefore, it must be re-built using one of the methods of combining the maximum vibration response of each mode. The commonly used methods are the **Square Root of Sum of Squares** (SRSS) and the **Absolute Sum**. These methods will be described later.

We will expand on point **4**. Figure 17-2 shows a series of Single Degree of Freedom (SDOF) oscillators all attached to a common base. Their natural frequencies are different, changing from the lowest 3Hz to the highest 30Hz. All SDOF oscillators have the same damping ratio. This base is subjected to excitation with the acceleration time history shown in Figure 17-1.

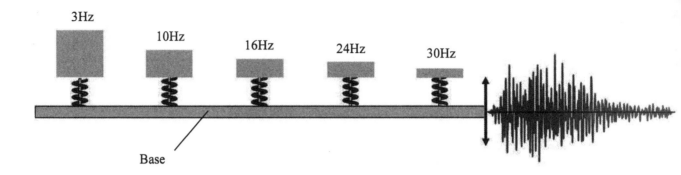

Figure 17-2: An illustrative explanation of the Response Spectrum concept: five SDOF oscillators are subjected to a base excitation with acceleration time history recorded during an earthquake. The natural frequency of each SDOF is indicated.

The natural frequencies of the above five oscillators define four ranges of frequencies: 3-10Hz, 10-16Hz, 16-24Hz, 24-30Hz.

A **Time Response** analysis of the set of SDOFs in Figure 17-2 is conducted and the maximum absolute displacement, velocity, and acceleration are found for each oscillator. The **Time Response** analysis is easy because of simplicity of this system; it has only five DOFs. The corresponding **Response Spectrum** curves will have five points. An example of an **Acceleration Response Spectrum** curve is shown in Figure 17-3.

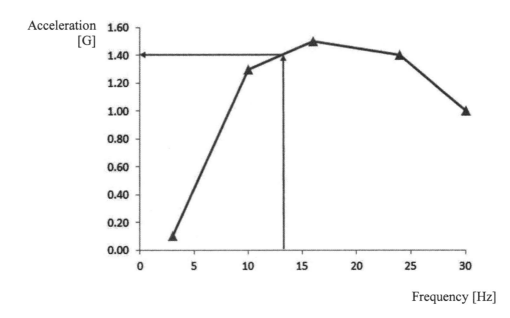

Figure 17-3: An Acceleration Response Spectrum curve generated from the results of a Time Response analysis of a series of SDOFs with frequencies 3Hz, 10Hz, 16Hz, 24Hz, and 30Hz.

The curve serves as an illustrative example only; it does not correspond to any seismic shock.

Response for frequencies falling in-between the natural frequencies of the oscillators shown in Figure 17-2 may be interpolated. Figure 17-3 shows a linear interpolation.

Assume for a moment that natural frequencies of a structure are dominated by a single mode and the direction of excitation is aligned with that mode. In this case the response (e.g. maximum acceleration) can be read from a graph. Finite Element Analysis programs merely automate interpolation and combine effects of multiple modes and different excitation directions.

The direction of excitation with regard to the direction of modal shape must be taken into consideration when a **Response Spectrum** curve is constructed. This is illustratively shown in Figure 17-4.

Figure 17-4: Illustrative explanation of the Response Spectrum concept taking into consideration that the direction of excitation and the directions of modal shapes may not be aligned.

Excitation is applied perpendicularly to the base; oscillators are located "at an angle" to the direction of excitation.

Most often the **Response Spectra** curves are produced by subjecting the series of SDOFs to a synthesized acceleration time history that does not correspond to any particular earthquake but represents an earthquake that may happen in a given geographical region. The synthesized **Response Spectra** curves can be found in seismic codes and are used as inputs to seismic analyses. An example of some **Response Spectra** for equipment installed in a building during a seismic event is shown in Figure 17-5.

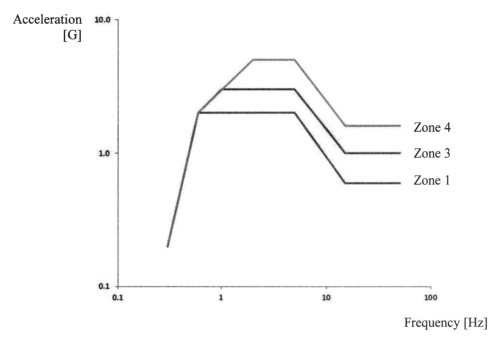

Figure 17-5: Sample Response Spectra for different seismic zones.

The **Response Spectra** shown in Figure 17-5 are not meant to be used as an input to seismic analysis of an entire building. These are accelerations in the horizontal direction of an upper floor of a building during an earthquake.

Even though the **Response Spectrum** method is ubiquitous in earthquake engineering, it has important limitations. Finite Element Analysis programs like **SOLIDWORKS Simulation** calculate the response of each mode independently and then must somehow combine them together. The maximum responses in different modes do not have to coincide in time; therefore, these responses cannot be combined directly. **SOLIDWORKS Simulation** offers a choice of four approximate methods of combining individual modal responses:

- Square Root Sum of Squares (SRSS)
- Absolute Sum (ABS)
- Complete Quadratic Combination (CQC)
- Naval Research Laboratory (NRL)

The **SRSS** estimates the peak response by the square root of the sum of the maximum responses squared. The **ABS** method assumes that the maximum modal responses occur at the same time for all modes. This is the most conservative method of modal combination. The **CQC** is based on random vibration concepts, and **NRL** takes the absolute value of the largest response and adds it to the **SRSS** response of other modes.

How many modes should be considered when modeling a structure's response to a seismic shock? Most often seismic analysis is conducted to demonstrate compliance with certain code requirements. That specific code will specify the minimum mass participation factor in the model used for seismic response. Hence, we need to use as many modes as necessary to ensure the required mass participation.

We'll demonstrate the use of **SOLIDWORKS Simulation** for the analysis of a seismic response using the assembly model titled FRAME, shown in Figure 17-6. Base excitation in the form of a **Response Spectrum** will be similar to the Zone 4 excitation in Figure 17-5. Our objective is to analyze displacements and accelerations experienced by the FRAME installed on an upper floor of a building experiencing **Zone 4** seismic shock. Sharp re-entrant edges in the model geometry are acceptable because stress analysis is not an objective of this exercise.

The model consists of a steel cylinder resting on a plastic frame as shown in Figure 17-6.

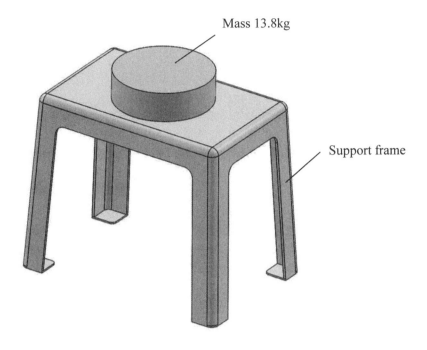

Mass 13.8kg

Support frame

Figure 17-6: The FRAME model used for the analysis of a seismic shock.

All legs have fixed restraints applied to the pads.

Notice sharp re-entrant edges in the model.

A **Response Spectrum** study does not allow for use of reference geometry in the definition of the direction of excitation. Excitation can be only applied in the global X, Y, or Z directions. Therefore, the assembly model is located "at an angle" with respect to the global coordinate system; excitation will be applied in the X direction (Figure 17-7).

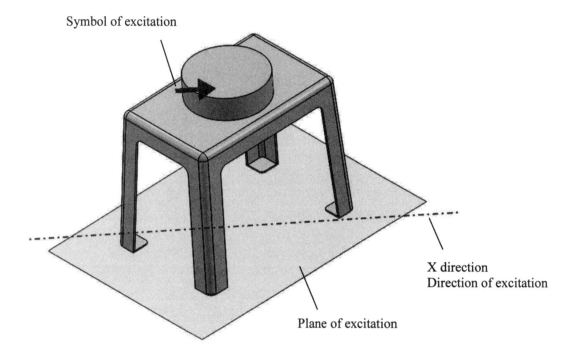

Symbol of excitation

X direction
Direction of excitation

Plane of excitation

Figure 17-7: Location of the FRAME relative to the global coordinate system of the assembly.

The excitation symbol is an arrow shown as crossing the center of mass of the assembly. The actual plane of excitation and the direction of excitation (aligned with X direction) are shown in this illustration.

Prior to the **Response Spectrum** analysis we'll find modes of vibration within the range of frequencies 0-50Hz. Create **Frequency** study *Modal* with default five modes and define **Fixed** restrains to four pads as shown on Figure 17-6. Mesh the assembly with a **Curvature base mesh** with the default element size.

Three modes within the range of 0-50Hz are shown in Figure 17-8.

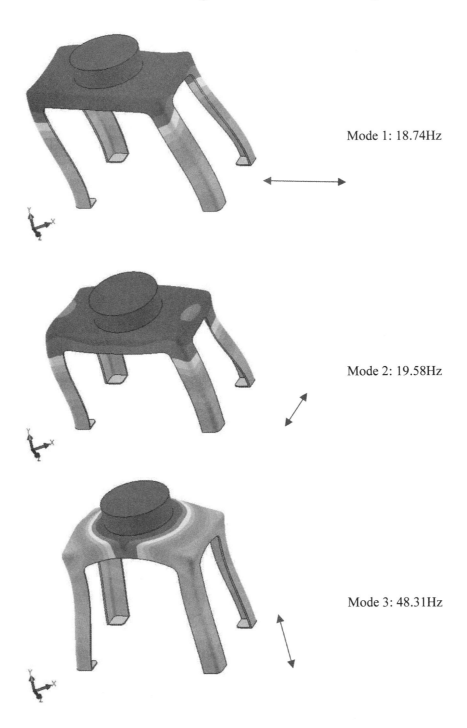

Mode 1: 18.74Hz

Mode 2: 19.58Hz

Mode 3: 48.31Hz

Figure 17-8: The three modes of vibration of the FRAME within the range of 0-50Hz. Arrows indicate the direction of vibration.

Mode 3 is aligned with Y axis of the global coordinate system.

The direction of base excitation is along the global X direction; therefore, mode 1 and mode 2 will participate in the vibration response. Mode 3 is orthogonal to the direction of excitation and will not be excited by the applied base excitation. The **Response Spectrum** method uses the **Modal Superposition Method** to find the vibration response; therefore, the vibration response will be a superposition of the responses of mode 1 and mode 2.

Create a **Response Spectrum** study as shown in Figure 17-9, call it *ABS*; we will use the ABS method to combine individual modal responses.

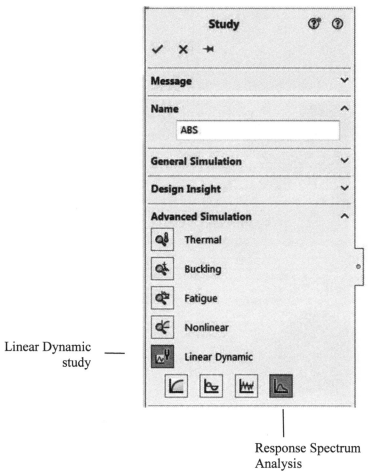

Linear Dynamic study

Response Spectrum Analysis

Figure 17-9: Definition of a Linear Dynamic study with the Response Spectrum Analysis option.

Define a **Uniform Base Excitation** as shown in Figure 17-10.

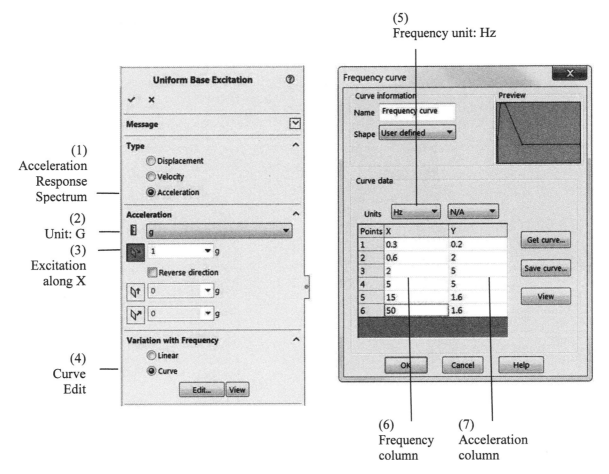

Figure 17-10: Definition of the base excitation.

Numbers in the acceleration column are multipliers to the gravitational acceleration entered in step 3. You may copy the Response Spectrum table from the spreadsheet titled FRAME.xlsx and paste it into the Frequency curve window.

Define restraints as shown in Figure 17-6 and mesh using a **Curvature based mesh** with default settings. The mesh is the same as the one used in the Frequency study; it is shown in Figure 17-11.

Figure 17-11: The FRAME assembly after meshing.

A curvature based mesh is recommended for easier control of turn angle on curved faces. Here, the minimum number of elements on circle is 8.

Define the study properties as shown in Figure 17-12.

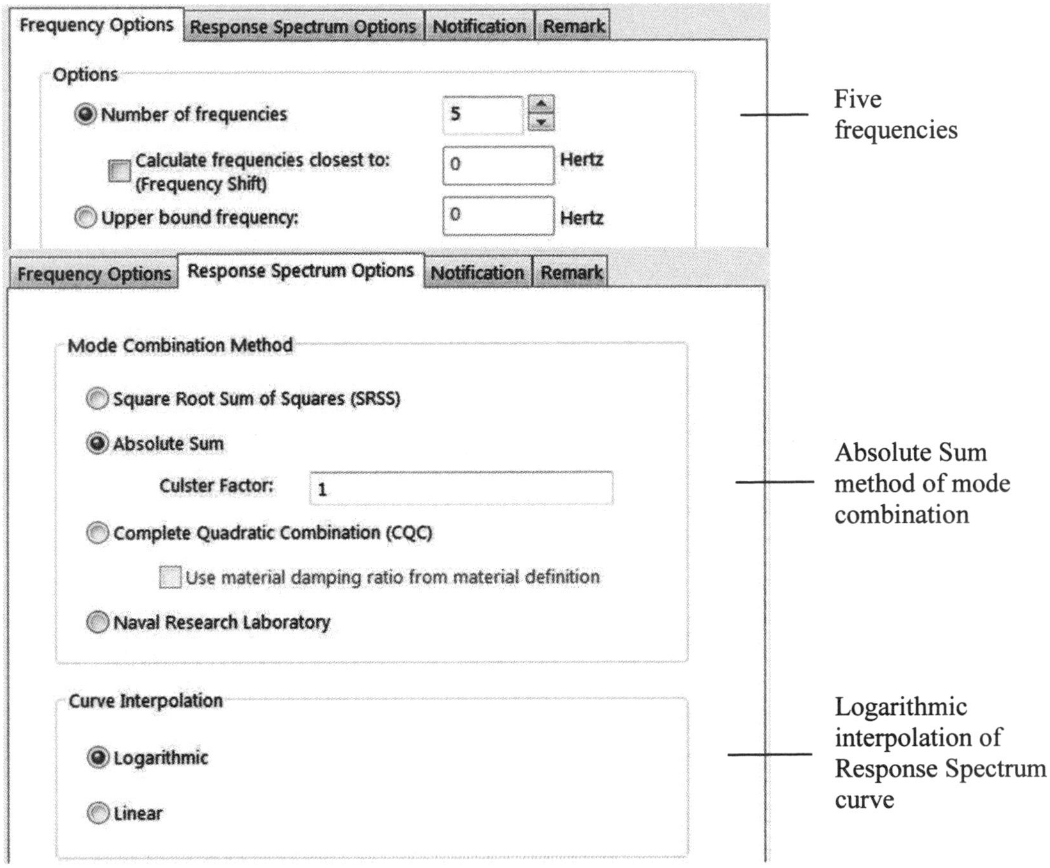

Figure 17-12: Response Spectrum Study properties: Frequency Options and Response Spectrum Options.

Since five frequencies are specified, the first five modes may possibly contribute to the seismic response.

In preparation for a **Response Spectrum** analysis, we conducted a modal analysis. The results of the modal analysis indicated that only the first two modes will participate in the seismic response. Yet, we specify five modes in **Frequency Options**. This is to prove that indeed only the first two modes are important in the seismic response; it will be clearly seen from the modal mass participation results.

Obtain the solution and create an acceleration plot as shown in Figure 17-13.

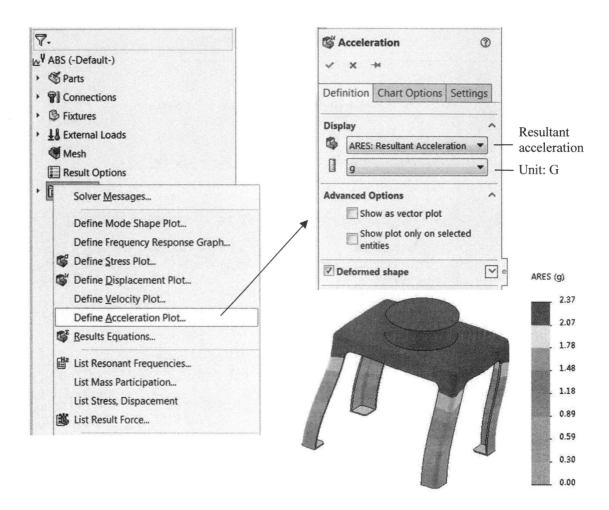

Figure 17-13: The maximum resultant acceleration during the Zone 4 seismic event.

The maximum resultant acceleration relative to the base is 2.37G.

Even though the maximum acceleration plot (as well as other plots) may be animated, the animation has no physical meaning in a **Response Spectrum** study. Remember that the results show the maximum values of acceleration during the entire duration of excitation. It is unknown at which point in time the maximum response happens. Also remember that the maximum response is the result of an arbitrary combination of individual modal responses, here performed using the **ABS** method. The same applies to all results of a **Response Spectrum** analysis.

Create the **Effective mass participation factor** window as shown in Figure 17-14.

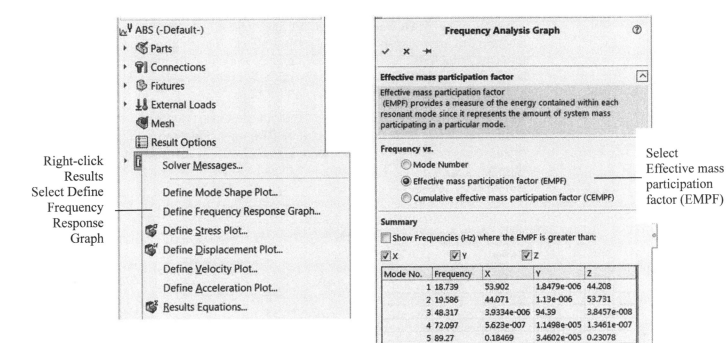

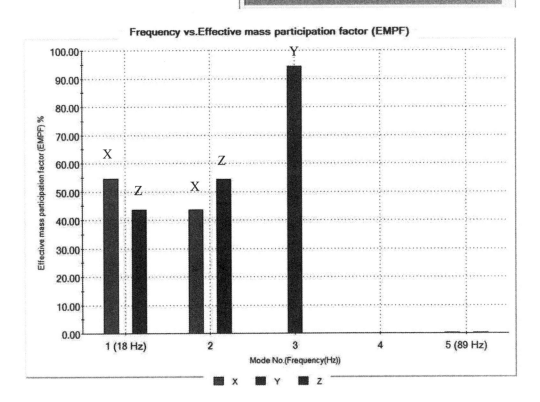

Figure 17-14: Mass participation of the first five modes; mode 4 and mode 5 do not participate in the response.

Directions X, Y, Z are identified by colors. In this black and white illustration they are identified by added labels X, Y, Z; review this graph on screen.

A review of the **Mass Participation** indicates that only the first and second modes participate in response to the seismic shock. The third mode with its frequency of 48.3Hz is still within the range of excitation frequencies 0-50Hz, but its direction is orthogonal to the direction of excitation. Modes 4 and 5 have negligible mass participation in all directions.

To gain a better understanding of how individual modes contribute to the seismic response, we will create two more **Response Spectrum** studies: *Seismic shock mode 1* and *Seismic shock mode 2* with **Frequency Options** shown in Figure 17-15. Each study will model a seismic response using only one mode.

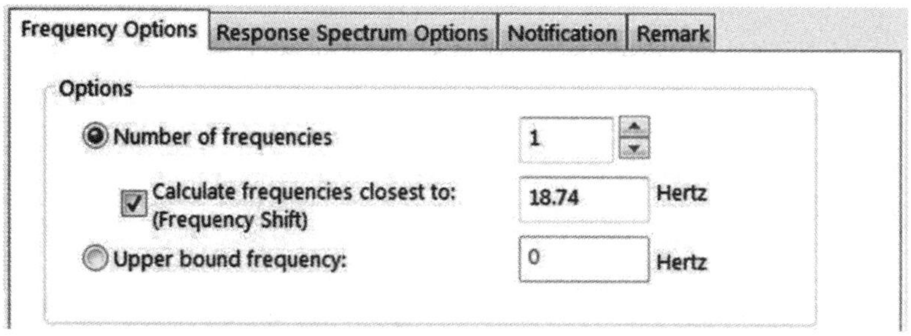

Frequency Options in study
Seismic shock mode 1

One mode
Frequency shift 18.74Hz

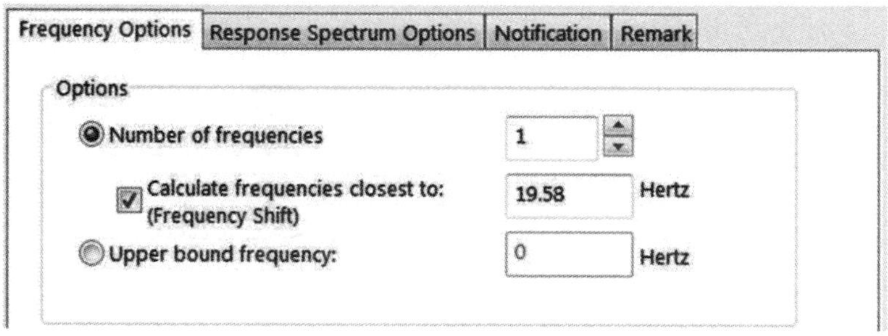

Frequency Options in study
Seismic shock mode 2

One mode
Frequency shift 19.58Hz

Figure 17-15: Frequency Options in the two studies used to analyze the individual mode contributions to the seismic response.

Seismic shock mode 1 study is based on the first mode: 18.74Hz.

Seismic shock mode 2 study is based on the second mode: 19.58Hz.

Notice that the **Frequency Options** of study *Seismic shock mode 1* could also be defined by specifying the number of frequencies equal to one. This is because 18.74Hz is the frequency of the first mode. The method of combining individual modal responses is irrelevant in *Seismic shock mode 1* and *Seismic shock mode 2* because each study is based on one mode only.

Results of the three studies: *ABS, Seismic shock mode 1* and *Seismic Shock mode 2* are shown in Figure 17-16:

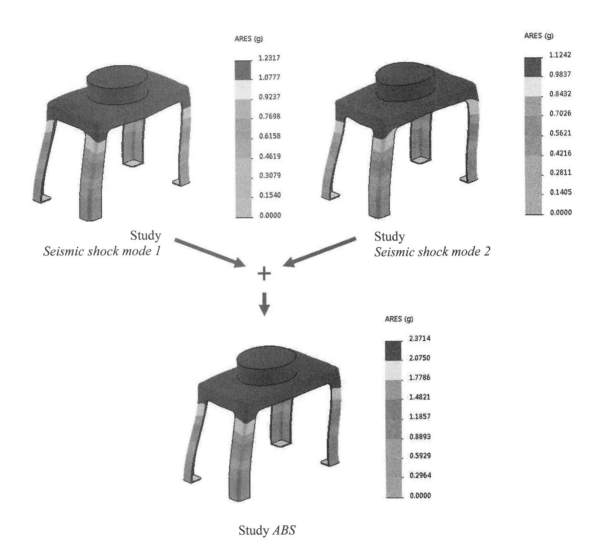

Figure 17-16: Acceleration results of studies based on mode 1 and mode 2 (top), and acceleration results based on the five modes out of which only the first two contribute to the response (bottom).

The maximum acceleration based on mode 1 is 1.23G; the maximum acceleration based on mode 2 is 1.12G. The maximum acceleration based on both modes is 2.37G.

The maximum acceleration found in study *ABS* is the **Absolute Sum** of the maximum acceleration in each mode. The difference in the second decimal place is caused by numerical error.

Run three more studies to try other **Mode Combination Methods** available in SOLIDWORKS Simulation:

Study *SRSS*: **Square Root of Squares (SRSS)**

Study *CQC*: **Complete Quadratic Combination (CQC)**

Study *NRL*: **Naval Research Laboratory**.

The summary of results is shown in Figure 17-17.

Square Root Sum of Squares

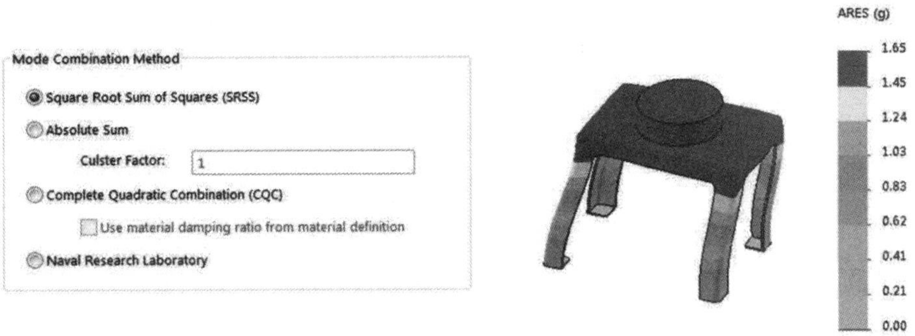

Absolute Sum

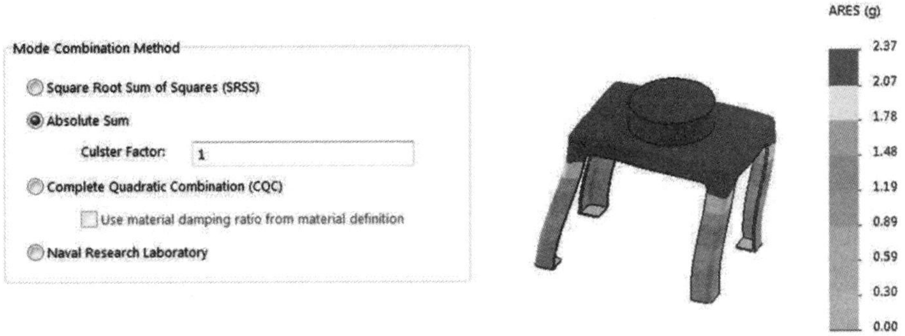

Complete Quadratic Combination

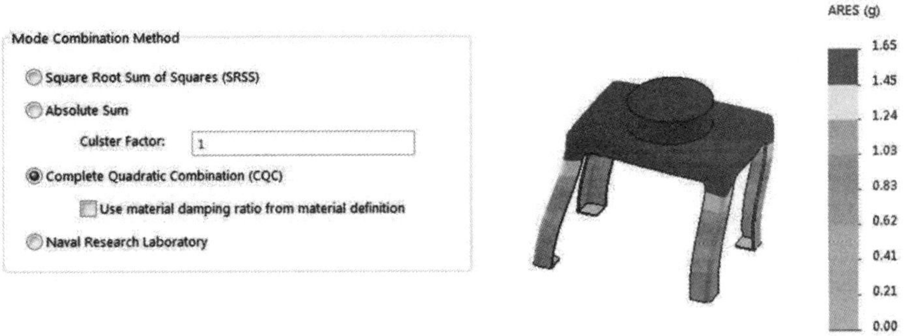

Naval Research Laboratory

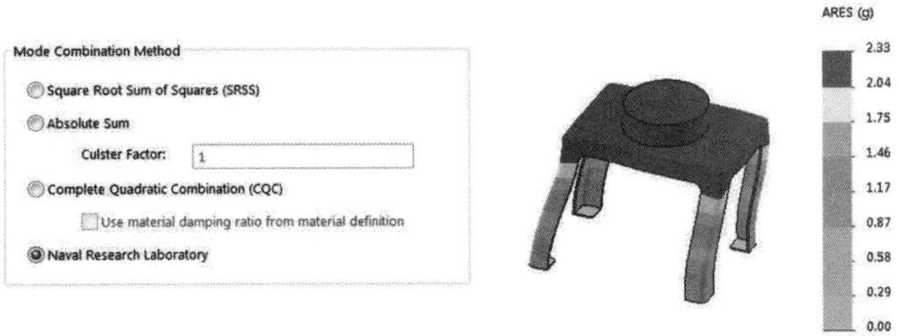

Figure 17-17: Acceleration results of study Seismic shock (based on two modes) using different methods of mode combination.

The Absolute Sum is the most conservative method.

The method of modes combination used in seismic analysis is usually based on specific code requirements that we have to meet. The maximum displacements and the maximum stress results using the SRSS method are shown in Figure 17-18.

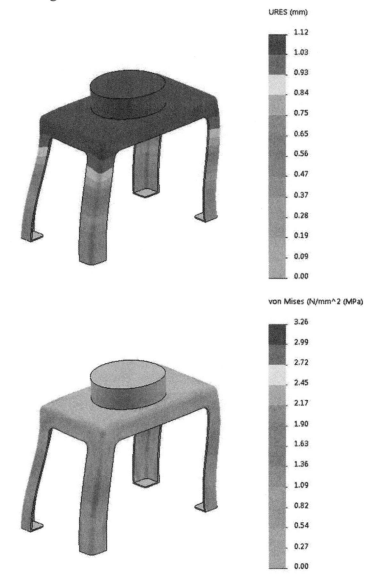

Figure 17-18: The maximum displacement (top) and the maximum von Mises stress during the seismic shock.

The SRSS method of mode combination is used.

The pass/fail result of seismic simulation is often made by comparing these results to requirements of the applicable seismic code.

To introduce the Response Spectrum Analysis we used a simplified model with sharp re-entrant edges causing stress singularities. If the code specified the maximum allowable stress, the model would have to be modified to avoid stress singularities.

Summary of studies completed

Model	Configuration	Study Name	Study Type
FRAME.sldasm	*Default*	*Modal*	Frequency
		ABS	Response Spectrum Analysis
		Seismic shock mode 1	Response Spectrum Analysis
		Seismic shock mode 2	Response Spectrum Analysis
		SRSS	Response Spectrum Analysis
		CQC	Response Spectrum Analysis
		NRL	Response Spectrum Analysis

Figure 17-19: Names and types of studies completed in this chapter.

18: Nonlinear vibration

Topics covered

- ❑ Differences between linear and nonlinear structural analysis
- ❑ Types of nonlinearities
- ❑ Bending stiffness
- ❑ Membrane stiffness
- ❑ Modal damping
- ❑ Rayleigh damping
- ❑ Linear Time response analysis
- ❑ Nonlinear Time response analysis
- ❑ Modal Superposition Method
- ❑ Direct Integration Method

Difference between linear and nonlinear structural analysis

A structural analysis problem, be it static or time dependent, is always concerned with stiffness of the analyzed structure. If stiffness remains constant during the process of load application then the problem is linear. If stiffness changes then the problem is nonlinear. Nonlinear analyses are classified into different types based on the origin of the nonlinear behavior. For example, in nonlinear geometry analysis stiffness changes because of changes to a structure's geometry, in nonlinear material analysis stiffness changes because of changes in material properties etc. There are many other types of nonlinear analyses. For more information refer to "**Engineering Analysis with SOLIDWORKS Simulation.**"

The nonlinearity in the problems analyzed in this chapter is geometric; it is caused by changes in the structure's shape and requires **Large displacement** analysis. However, the term **Large displacement** is confusing because it implies that displacements must be large in order for the nonlinear effects to show up even though in many cases small displacements are sufficient to change the stiffness significantly.

Bending and membrane stiffness

Consider a beam represented by the model PLANK, supported on both sides by hinges and subjected to 1MPa pressure, as shown in Figure 18-1. The beam material is Alloy Steel, with a thickness of 25.4mm (1").

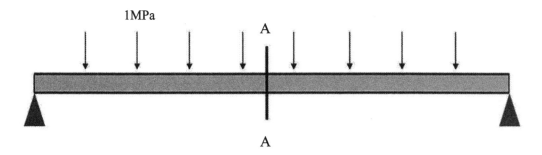

1MPa

A

A

Figure 18-1: A flat beam simply supported on both ends.

The two hinges supporting the beam can't move in any direction.

The stress distribution across the beam thickness in the cross section A-A changes as the beam deformation progresses (Figure 18-2).

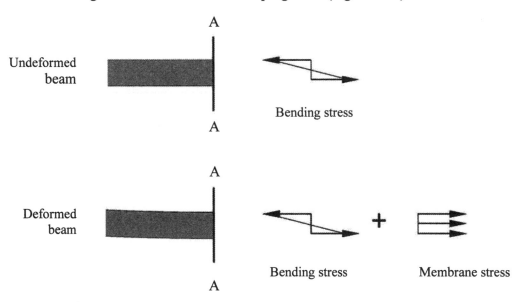

Undeformed beam

Bending stress

Deformed beam

Bending stress   Membrane stress

Figure 18-2: Stress distribution in cross-section A-A in the undeformed beam (top) and in the deformed beam (bottom).

The flat beam responds to load only with bending stress. The curved beam responds to load with bending stress and with tensile stress called membrane stress. Notice that membrane stresses wouldn't come into existence if one of the supports shown in Figure 18-1 could slide in the horizontal direction.

When the load is first applied and the beam is still flat, the beam can respond only with bending stresses. We may say it only has bending stiffness. As deformation progresses and the beam becomes curved, tensile stresses develop and the beam acquires a new type of stiffness: membrane stiffness. The name "membrane" comes from the analysis of thin walls called membranes where bending stiffness is negligible and all stiffness is generated by membrane stress. In our case, the stiffness of the curved beam is a superposition of bending and membrane stiffness.

The PLANK model has two configurations: *01 solid* and *02 surface*. We'll use *02 surface* which allows for meshing with shell elements and faster solution.

We will use static analysis to demonstrate the nonlinearity. A static analysis accounting for nonlinear effects requires either a **Static** study with the **Large displacements** option, or a **Nonlinear** study with the **Large displacements** option. We'll use **Static** studies.

Set up and run two **Static** studies: a linear analysis titled *01 static LIN* and a nonlinear static analysis titled *02 static NL*. To create the nonlinear analysis *02 static NL* you may copy *01 static LIN* into *02 static NL* and select **Large displacements** in the study properties. Remember to define restraints as **Immovable** (not **Fixed**) to model hinge supports on both sides. Contrary to solid elements, this is because shell elements differentiate between these two types of restraints.

Figure 18-3 shows the comparison of displacement results obtained with a **Static** study with and without the **Large displacement** option.

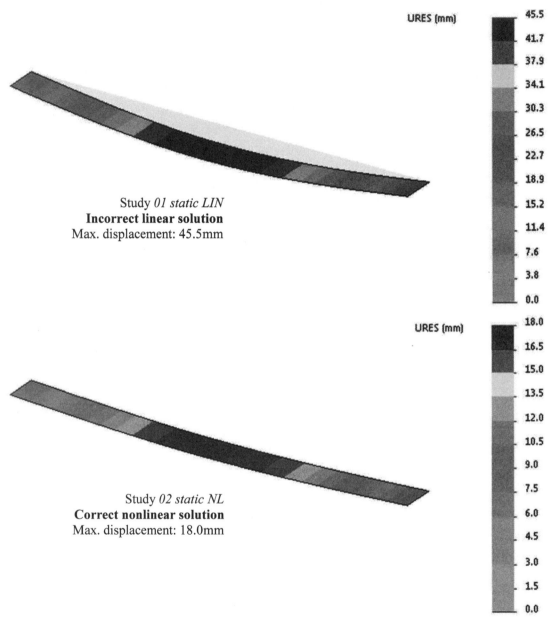

Study *01 static LIN*
Incorrect linear solution
Max. displacement: 45.5mm

Study *02 static NL*
Correct nonlinear solution
Max. displacement: 18.0mm

Figure 18-3: Comparison of resultant displacement results of the PLANK treated as linear (top) and nonlinear (bottom) problem.

The scale of deformation (1:1) is the same in both plots.

As can be seen in Figure 18-3, the linear solution ignores the stiffening effect of membrane stresses. Consequently, it severely underestimates the stiffness. Having reviewed displacement results, compare von Mises stress results and notice that the linear analysis produces stresses above yield and the nonlinear analysis produces stresses below yield (Figure 18-4).

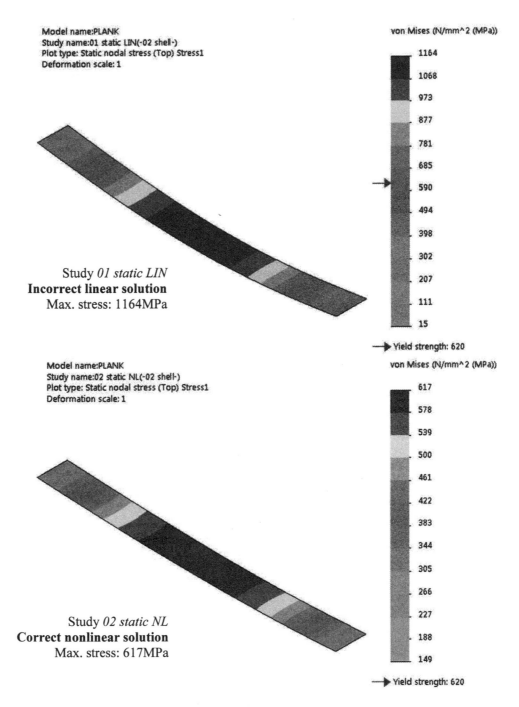

<u>Figure 18-4: Comparison of von Mises stress results of the PLANK treated as linear (top) and nonlinear (bottom).</u>

The scale of deformation (1:1) is the same in both plots. Stress results on the Top of shell elements are shown. The top of elements is the side opposite to where pressure is applied. Review the results on the bottom side. Notice that the linear analysis produces the same von Mises stress on both sides while the nonlinear analysis produces a much higher stress on the top side. This is the effect of membrane stress present in nonlinear analysis.

After this short review demonstrating the nonlinear nature of the PLANK problem, we move on to the analysis of vibration which will include nonlinear vibration response. This type of analysis is computationally demanding; this is why we use the PLANK model in the *02 surface* configuration suitable for meshing with shell elements. The use of shell elements reduces the solution time significantly. To reduce the solution time even more, use shell elements with a size of 20mm. Large elements are acceptable because the stress distribution across the shell element thickness is built into the element definition and is not dependent on the element size. Furthermore, the overall model stiffness is weakly dependent on the element size. You may want to confirm this by running the same problem with several meshes using different element sizes.

Set up a **Frequency** study titled *03 modal* to find the first two modes of vibration. Again, remember to define **Immovable** (not **Fixed**) restraints. Results of the modal analysis are shown in Figure 18-5.

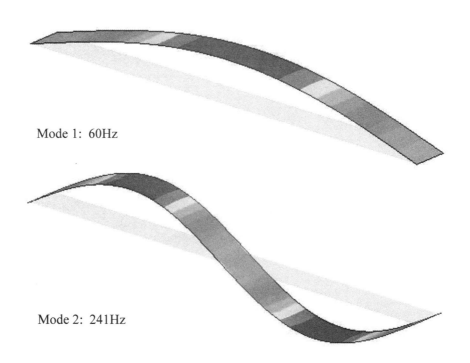

Mode 1: 60Hz

Mode 2: 241Hz

Figure 18-5: Results of modal analysis.

The first two modes are shown. The rotation on both ends of the plank is enabled by hinge restraints. Undeformed shape is superimposed on the modal shapes.

Notice that modal analysis works with the initial stiffness; it does not take into consideration the membrane stiffening effect.

Having found the modes of vibration, we proceed with a linear **Time Response** analysis available in a **Linear Dynamic** study with the **Modal Time History** option. We already know that the problem is nonlinear. The results of the linear **Time Response** will serve only as a comparison to the results of the subsequent nonlinear **Time Response** analysis.

In the **Modal Time History** study, the PLANK model will be subjected to an impulse pressure shown in Figure 18-6.

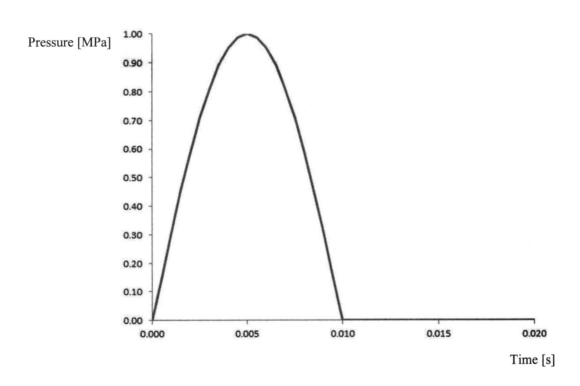

Figure 18-6: Time history of pressure acting on the top face of the PLANK.

During the first 0.01s, pressure rises from zero to 1MPa and drops back to zero following a sine curve with frequency of 50Hz. After 0.01s pressure remains zero until the end of the analysis.

Define a **Modal Time History** study titled *04 dynamic LIN* with properties shown in Figure 18-7.

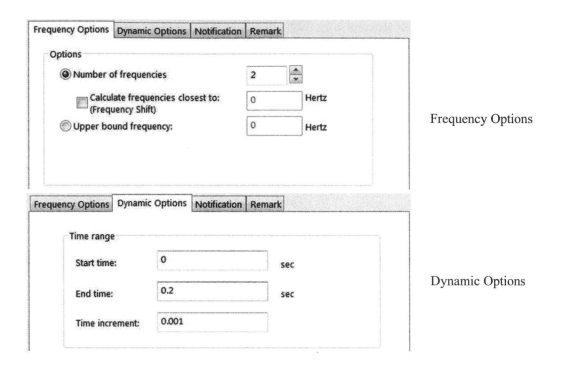

Figure 18-7: Properties of the Modal Time History study.

The study is based on two modes; its duration is 0.2s with a time step of 0.001s.

Define modal damping individually for each mode (Figure 18-8).

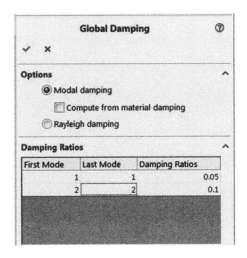

Mode 1: damping 5%
Mode 2: damping 10%

Figure 18-8: Modal damping definition.

Modal damping is defined individually for each mode. We assume these values are known from testing.

Following the definition of the study properties (Figure 18-7), the **Modal Time History** study is based on the superposition of the responses of two modes, each one with different modal damping (Figure 18-8).

Apply a 1MPa pressure to the top face (Figure 18-1); the same as in a static analysis and make it time dependent as shown in Figure 18-9.

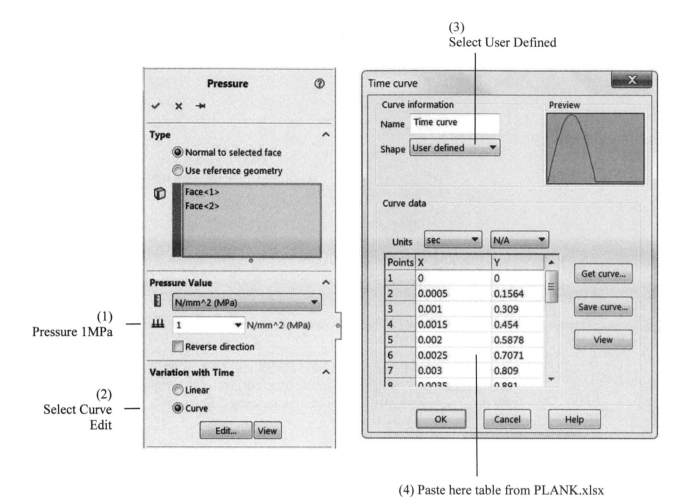

Figure 18-9: Definition of time dependent load.

Copy the table from the spreadsheet titled "PLANK.xlsx" into the Time Curve definition window. The table lists the time dependent multiplier to the specified pressure magnitude 1MPa.

Notice that the time duration in the table copied from PLANK.xlsx is only 0.02s, the same as shown in Figure 18-6. The analysis duration specified in the study properties shown in Figure 18-6 is 0.2s. The program will extrapolate the time curve as zero from 0.02s to 0.2s.

Define **Immovable** restraints identical to the previously completed studies.
Define **Results Options** as shown in Figure 18-10.

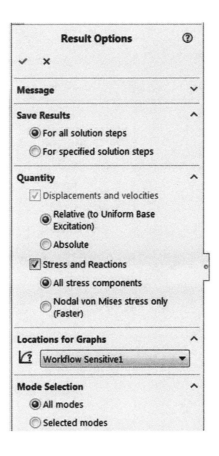

Figure 18-10: Definition of Result Options.

Select the Workflow Sensitive1 sensor which has been defined in the SOLIDWORKS model.

The study is now ready to run. Obtain the solution and notice that it solves very quickly because it is a linear analysis, and as such, it is based on the **Modal Superposition Method**.

Construct a graph of displacement time history in the mid-span, where the **Workflow Sensitive** sensor has been defined (Figure 18-11).

Workflow
Sensitive sensor

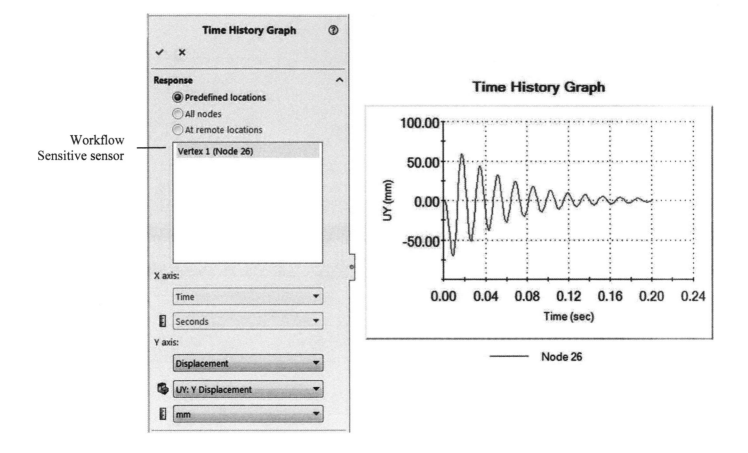

Figure 18-11: Displacement time history in the middle of the beam span.

Select the Workflow Sensitive sensor which has been defined in the SOLIDWORKS model; select the UY displacement component.

Time history of in-plane stress SX will be discussed together with the results of nonlinear analysis.

Notice that when the impulse load disappears after 0.01s the beam performs free damped vibration behaving as a single degree of freedom oscillator. Even though the **Modal Time History** study is based on two modes, the first mode dominates the vibration response.

Previously completed static analysis of this model indicated the importance of nonlinear effects. The nonlinear effects are also present in the vibration response. As we'll soon demonstrate, the above linear solution is incorrect.

We will now perform analysis of the nonlinear vibration response starting with a comparison of the number of Degrees of Freedom (DOF) in the linear and nonlinear vibration problems. The linear vibration problem uses the **Modal Superposition Method**. Two modes have been specified in the study properties; therefore, the problem is represented by two DOFs. The nonlinear problem can't use the **Modal Superposition Method** and must solve all equations of motion in terms of all degrees of freedom present in the model using the **Direct Integration** method.

Look up the **Solver Message** in any of the completed studies to see that the number of DOFs in the model is 6408 (Figure 18-12).

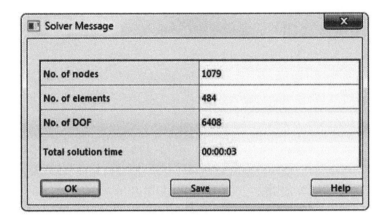

Figure 18-12: The number of DOFs in the PLANK model.

The number of DOFs in the model is 6408. It is not equal to the number of nodes multiplied by six because some DOFs are removed by the restraints.

The comparison between 2 DOFs in the linear problem and the 6408 DOFs in the nonlinear problem gives us an idea how numerically intensive a nonlinear vibration problem will be. This is why the PLANK model is so simple; otherwise it wouldn't solve reasonably fast.

Another consequence of the inability to use the **Modal Superposition Method** is that we can't use modal damping. Damping in nonlinear problems must be defined as **Rayleigh damping**. **Rayleigh damping** makes an arbitrary assumption that the damping matrix is a linear combination of the mass and stiffness matrices. This assumption is a mathematical convenience for the purpose of simplification since there is no physical justification for this. **Rayleigh Damping** is specified by two damping constants: α and β which are used as multipliers of the mass matrix M and the stiffness matrix K when calculating damping matrix C:

$$[C] = \alpha[M] + \beta[K] \qquad \text{(Eq. 18-1)}$$

$$\frac{\alpha}{2\omega} + \frac{\beta\omega}{2} = \zeta \qquad \text{(Eq. 18-2)}$$

where ω is the frequency and ζ is the damping ratio.

Alpha damping α is a viscous damping component also known as mass damping. It characterizes damping of lower frequencies.

Beta damping β is a hysteresis damping component also known as solid or stiffness damping. It characterizes damping of higher frequencies.

Rayleigh damping components α and β can be found if two modes and their modal damping are known. In our case:

	Frequency Hz	Frequency rad/s	Modal damping %
Mode 1	60	378	5
Mode 2	241	1513	10

Figure 18-13: Modal frequencies and their damping ratios in the PLANK model.

We need to know the above values to find Rayleigh damping.

Using the data in the above table, α and β can be found by solving two equations:

$$\frac{\alpha}{2\omega_1} + \frac{\beta\omega_1}{2} = \zeta_1 \qquad \frac{\alpha}{2\omega_2} + \frac{\beta\omega_2}{2} = \zeta_2 \qquad \text{(Eq. 18-3)}$$

You may also use the spreadsheet "PLANK PAYLEIGH.xlsx" to find α = 20.17 and β = 0.00012.

Knowing α and β, equation (18-2) may be plotted to visualize how damping coefficients change with frequency.

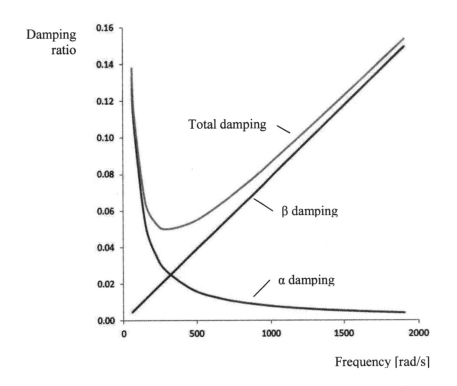

Figure 18-14: Damping ratio and its components as a function of frequency.

α damping drops with frequency; β damping is a linear function of frequency.

The graph in Figure 18-14 serves only to show the importance of **Alpha Damping** and **Beta Damping** for different frequencies; the total damping is not used in solving nonlinear vibration problems. The components α and β are used to formulate damping matrix C, as shown in equation (18-1).

Create a **Nonlinear** study titled *05 dynamic NL* with the **Dynamic** option (Figure 18-15):

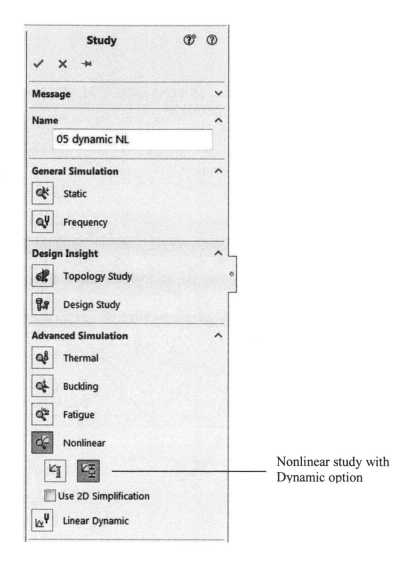

Nonlinear study with Dynamic option

Figure 18-15: Definition of a Nonlinear study with the Dynamic option.

Define damping as shown in Figure 18-16:

Figure 18-16: Definition of damping in a nonlinear vibration problem.

Enter the previously calculated constants α and β.

Apply the pressure load as a function of time the same way as defined in the linear study. Apply the restraints and use the mesh size the same as well.

We are now ready to define the properties of the nonlinear vibration study as shown in Figure 18-17:

End time 0.2s ——

Auto stepping with ——
initial time increment
0.002s

Large displacement ——

Update load direction

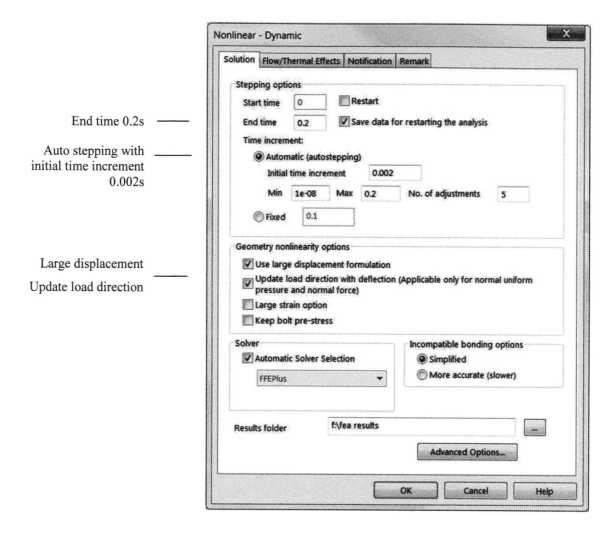

Figure 18-17: Properties of the nonlinear vibration study.

Specify the analysis duration as 0.2s; accept the default auto stepping; select Large displacement and Update load direction.

Large Displacements must be specified or else the problem would not be nonlinear and membrane stiffening would not be accounted for.

Run the solution and acknowledge the warning message produced by the nonlinear solver (Figure 18-18).

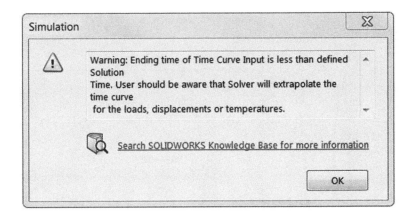

Figure 18-18: A warning message of the nonlinear solver issued when the duration of load is shorter than the duration of analysis.

The load duration is 0.01s, analysis duration is 0.2s. The end value of load is zero; therefore, the extrapolated load will be zero for the remaining 0.18s.

Obtain the nonlinear solution and construct displacement and stress graphs the same way as in the linear study.

Displacement responses for the linear and nonlinear solution are summarized in Figure 18-19.

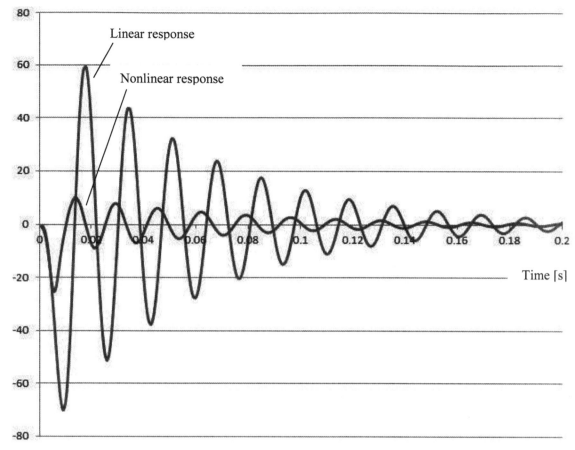

Figure 18-19: Comparison of linear and nonlinear UY displacement results in the mid-span of the PLANK.

The linear response doesn't account for membrane stiffening and produces displacement results with ~300% error.

The comparison between SX stress in the linear and nonlinear solutions in the sensor location is shown in Figure 18-20.

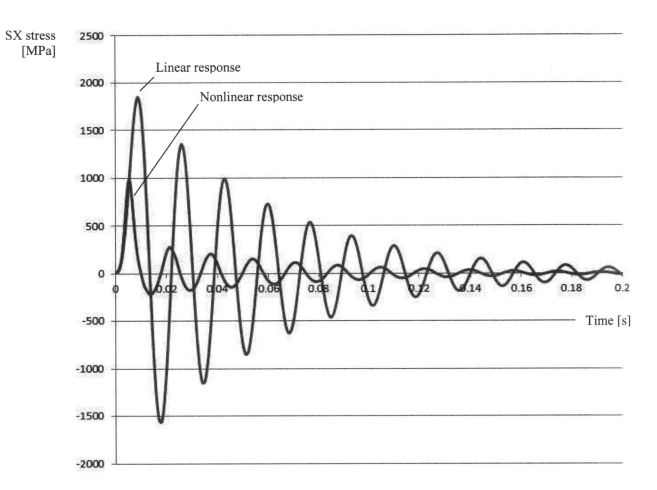

Figure 18-20: Comparison between linear and nonlinear SX stress results in the mid-span of the PLANK.

In the nonlinear response, the magnitude of tensile stress (positive) is larger than the magnitude of compressive stress (negative). This is due to the fact that the sensor location is on the top of the shell elements and does not change location from top to bottom with the reversal of beam deformation.

The frequency of the load time history (50Hz) is close to the first natural frequency of the model; this is the reason for high displacement and stress in the first cycle. Notice that SX stress exceeds the yield strength of the PLANK material (620MPa) in the first vibration cycle indicating that PLANK will experience yield. Whether or not this will result in a structural failure would require an analysis that includes a nonlinear material model.

All results presented above are reported at the sensor location shown in Figure 18-21.

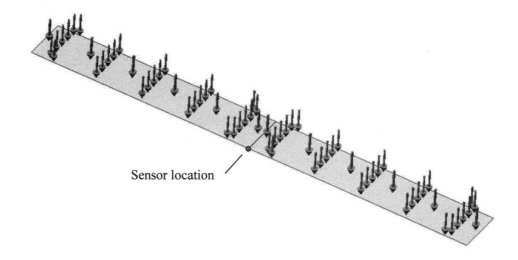

Figure 18-21: Sensor definition and its importance in the interpretation of results.

The sensor is defined at the end of a split line in the mid-span on the PLANK. Pressure symbols are shown; Immovable restraints symbols are hidden.

Review the shell element orientation (bottom is orange) to confirm that the top of the PLANK (the side where pressure is applied) has been meshed with bottoms of shell elements (Figure 18-22).

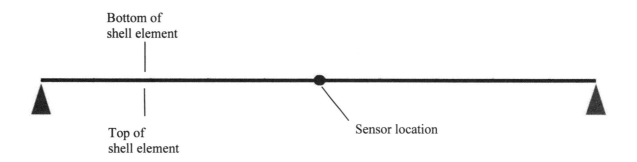

Figure 18-22: The sensor defined for the shells refers to the top of the shell element. The shell element model has no physical thickness.

The sensor does not differentiate between top and bottom of the surface.

The sensor reads stress at the top of elements (which is the bottom of PLANK) and its location does not change with the reversal of beam deformation. This explains the different magnitudes of tensile and compressive stress in the nonlinear SX time response observed in Figure 18-20 at the start of vibration when the effect of membrane stress is most noticeable.

Summary of studies completed

Model	Configuration	Study Name	Study Type
PLANK.sldprt	02 surface	01 static LIN	Static
		02 static NL	Static
		03 modal	Frequency
		04 dynamic LIN	Modal Time History
		05 dynamic NL	Nonlinear with Dynamic Option

Figure 18-23: Names and types of studies completed in this chapter.

19: Vibration benchmarks

A benchmark is a standard point of reference against which things may be compared or assessed. In vibration analysis these benchmarks are known analytical solutions. Solutions of commercial FEA programs may be compared to those known solutions to assess a program's performance.

A large number of vibration benchmarks are available. In this chapter we solve several NAFEMS benchmarks to see how well our results match those benchmark targets.

Since you are already well familiar with all types of vibration analyses with **SOLIDWORKS Simulation,** we'll take shortcuts discussing the setup of these benchmark problems. In some cases the same model will be used for more than one benchmark test and studies will be created in different model configurations.

We present a summary of all studies now rather than at the end of the chapter.

Model	Configuration	Study name	Study type
NAFEMS TEST FV2	01 solids	01 solids	Frequency
	02 shells	02 shells	Frequency
	03 beams	03 beams	Frequency
NAFEMS TEST 5	Default	TEST 5	Frequency
		TEST 5H	Harmonic
		TEST 5T LIN	Modal Time History
		TEST 5T NL	Nonlinear Dynamic
NAFEMS TEST 13	Default	TEST 13	Frequency
	Default	TEST 13R	Random

Figure 19-1: Summary of NAFEMS Benchmark tests, corresponding SOLIDWORKS models, model configurations, study names and study types.

NAFEMS test FV2

Pin-ended double cross: in-plane vibration.

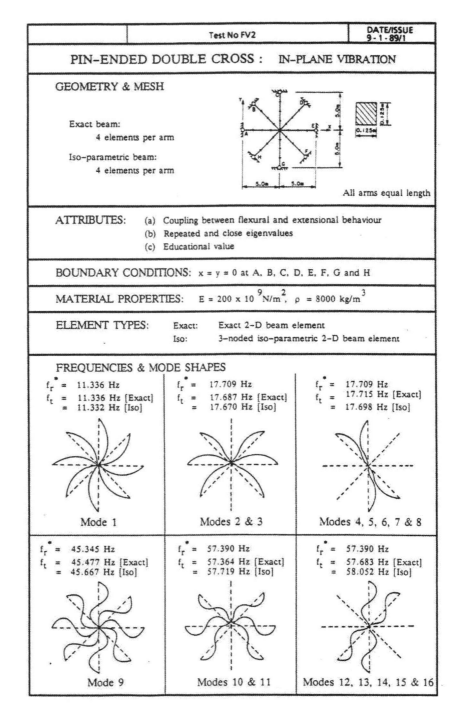

Figure 19-2: NAFEMS benchmark test FV2.

Material properties require custom material definition:

E=2e11Pa, ρ=8000kg/m³.

Open part model NAFEMS TEST FV2 and review its geometry: eight beams 5m long form a double cross; the beam cross section is a square with side length 0.125m. Material property: E=200000MPa, ν=0.3, ρ = 8000kg/m³. All beams are pin supported at the ends (Figure 19-3).

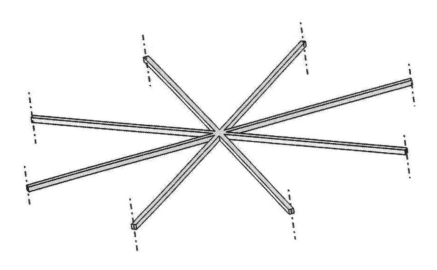

Figure 19-3: Geometry of NAFEMS FV2 *model* shown in the *01solids* configuration.

Pin supports allow for rotation of each end about its own axis. The axes are shown as dotted lines.

Notice that the model geometry lends itself to different representations: volume (solid) geometry for meshing with solid elements, surface geometry for meshing with shell elements and curve (wireframe) geometry for meshing with beam elements. You'll find these three representations in configurations *01 solids*, *02 shells*, and *03 beams*. All configurations use the same custom material with properties specified in the FV2 test.

The FV2 test specifies the use of beam elements; we'll expand it to include solid, shell and beam elements but we'll limit it to the first mode only. Therefore, we'll apply proper restraints to make sure that the first mode of vibration is in-plane and corresponds to what is required in FV2 benchmark test.

In each configuration create a **Frequency** study with the same name as the configuration name: *01 solids, 02 shells, 03 beams*.

Change to study *01 solids*, create a **Frequency** study and define the restraints for the solid element model as shown in Figure 19-4.

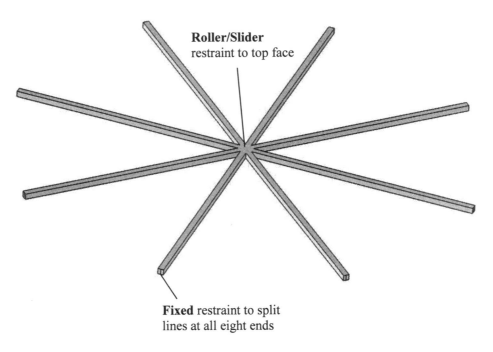

Roller/Slider restraint to top face

Fixed restraint to split lines at all eight ends

Figure 19-4: Restraints to the model in the *01 solids* study.

Restraints symbols are not shown. Roller/slide restraint may be applied to either the top face, to the bottom face or to both.

Remember that nodes of solid elements have three degrees of freedom and can't generate moment reactions. Therefore, **Fixed** restraints to a straight line produce hinges. The **Roller/Slider** restraint eliminates out of plane modes to produce in-plane modes as required by the test.

Use the default element size to mesh this model and run the solution.

Change to the *02 shells* study and define restraints for the shell element model as shown in Figure 19-5. To apply pin supports, use **Immovable** restraints. To define restraints enforcing 2D vibration, apply **Use Reference Geometry** restraints.

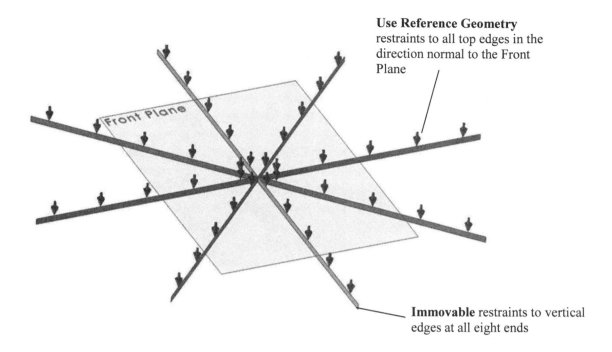

Use Reference Geometry restraints to all top edges in the direction normal to the Front Plane

Immovable restraints to vertical edges at all eight ends

Figure 19-5: Restraints to the model in the *02 shells* study.

The Use Reference Geometry support may be applied to the top edges or the bottom edges or to both. Immovable restraint symbols are not shown.

The **Immovable** restraints do not affect rotational degrees of freedom of the shell elements and allow for rotation. **Use Reference Geometry** restraints, use the **Front Plane** for reference, and eliminate translations in the direction normal to this plane. This is necessary to eliminate out of plane modes of vibration.

Define the shell element thickness as 125mm, use the default element size to mesh this model and run the solution.

Change to the *03 beams* study. The model geometry is visually identical to the model in the *03 solids* configuration, but the eight beams are not merged into one solid body; the model has eight solid bodies. In **Frequency** study change all solid bodies to beams (right-click each solid body and select **Treat as Beam**); calculate joints using **Edit Joints**; define restraints for beam elements as shown in Figure 19-6.

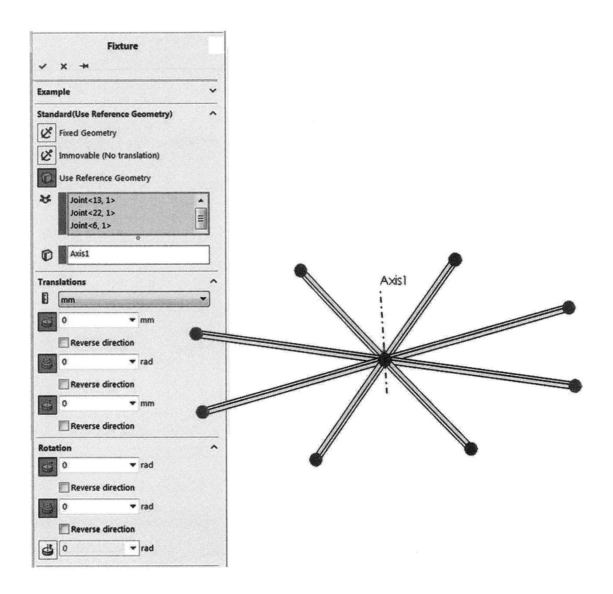

Figure 19-6: Restraints to the peripheral joints and to the central joint in the *03 beams* study.

Restraints to all nine joints are defined in a cylindrical coordinate system associated with Axis1. Restraint symbols are not shown.

Restraints applied to all joints (Figure 19-6) allow only rotation in the direction parallel to *Axis1*. This defines pin supports and eliminates lower out of plane modes. Higher out of plane modes are still possible, but for us they are irrelevant. All we want to find is the first mode of vibration to see how well the results of **SOLIDWORKS Simulation** match the benchmark target of 11.336Hz.

Results of all three studies are summarized in Figure 19-7.

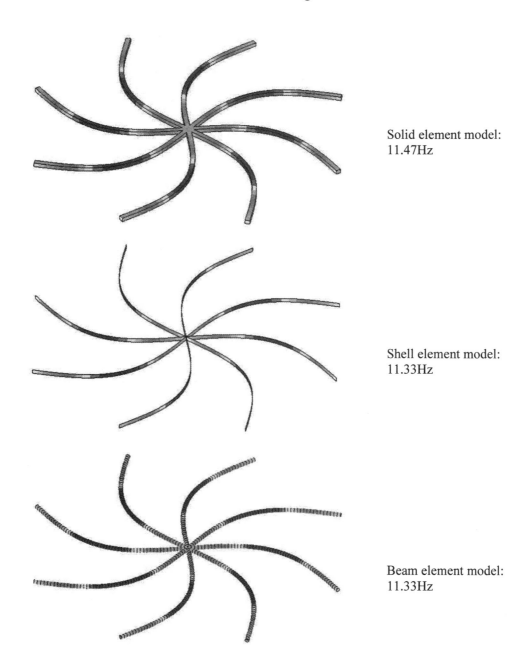

Solid element model:
11.47Hz

Shell element model:
11.33Hz

Beam element model:
11.33Hz

Figure 19-7: First mode of vibration of solid, shell, and beam element models.

All shapes correspond to the same first mode of vibration which is an in-plane mode. The solid element model reports marginally higher frequency because the arms do not connect exactly in the center and, for this reason, are shorter and have higher stiffness.

All results closely match the benchmark target solution of 11.336Hz.

NAFEMS test 5

Deep Simply-Supported Beam.

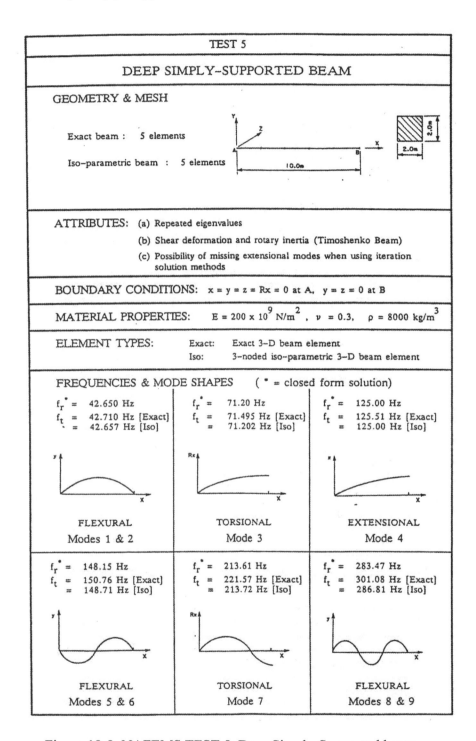

Figure 19-8: NAFEMS TEST 5; Deep Simply-Supported beam.

Custom material properties are the same as in test FV2.

Open model NAFEMS TEST 5 (Figure 19-9) and review its geometry: the beam is 10m long; the cross section is square with a side length of 2m. The beam has simple supports on both ends; rotation about its axis is prevented by restraint on one end; the material properties are E = 200000MPa, v=0.3, ρ = 8000kg/m³ as specified in Figure 19-8.

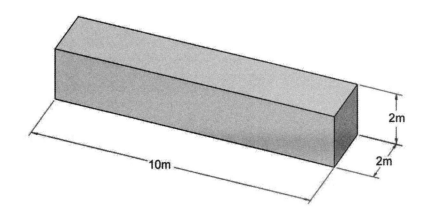

Figure 19-9: Deep Simply-Supported Beam, NAFEMS TEST 5.

The supports specified in TEST 5 would be difficult to model using a solid element model but easy using beam element model.

We'll test the model to see how well the results match the benchmark target frequencies.

Create a **Frequency** study *TEST* 5 and instruct **Simulation** to treat the solid geometry as a beam.

Define restraints for the beam element model as shown in Figure 19-10.

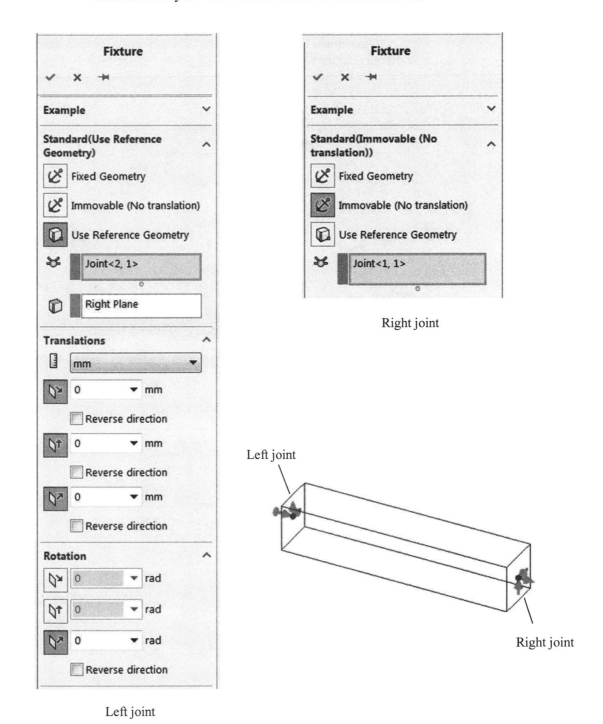

Right joint

Left joint

Right joint

Left joint

Figure 19-10: Restraints of the model in the *TEST 5* study.

Both joints have all translations removed. Additionally, the left joint has rotation about the beam axis removed.

Obtain solution of *TEST 5* study and review the modal shapes; the first mode is shown in Figure 19-11.

Model name:NAFEMS TEST 5
Study name:TEST 5(-Default-)
Plot type: Frequency Amplitude1
Mode Shape : 1 Value =   43.923 Hz
Deformation scale: 479.892

Figure 19-11: First mode of vibration of the beam element model; the frequency is 43.9Hz.

The second mode is identical in shape but rotated about the axis of the beam.

The result shown in Figure 19-11 closely matches the benchmark target solution of 42.710Hz. Review higher modes to see that the modal solution misses the extensional mode 4.

NAFEMS TEST 5H

Deep Simply-Supported Beam; Harmonic Forced Vibration Response.

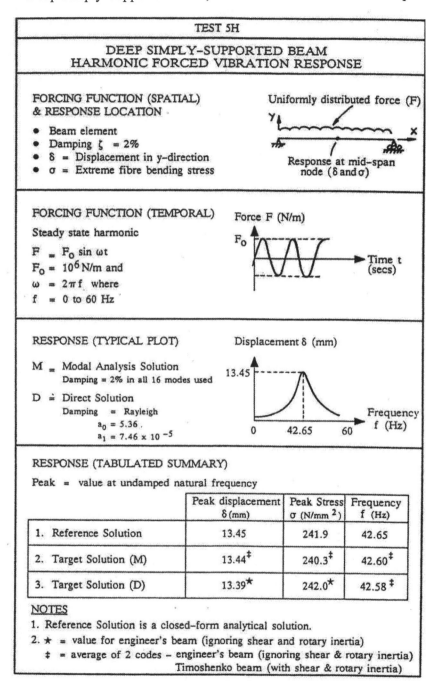

Figure 19-12: NAFEMS TEST 5H.

Geometry, material properties and restraints are the same as in NAFEMS TEST 5.

This test uses the same **SOLIDWORKS** NAFEMS TEST 5 model. The model is subjected to a harmonic excitation by a uniformly distributed force:

$$F = 10^6 * sin\omega t$$

$$\omega = 2\Pi f \quad 0 < f < 60Hz$$

The frequency of excitation changes from 0 to 60Hz.

Modal Damping is 2% in all 16 modes to be used for analysis. **Rayleigh Damping may** be used alternatively to **Modal Damping**; the constants are $\alpha=5.36$, $\beta =7.46e10^{-5}$.

The test target is the peak displacement in the mid-span of 13.45mm, and the peak stress is 241.9MPa for the excitation frequency of 42.65Hz.

Create **Harmonic** study *TEST 5H*.

Define study properties as shown in Figure 19-13.

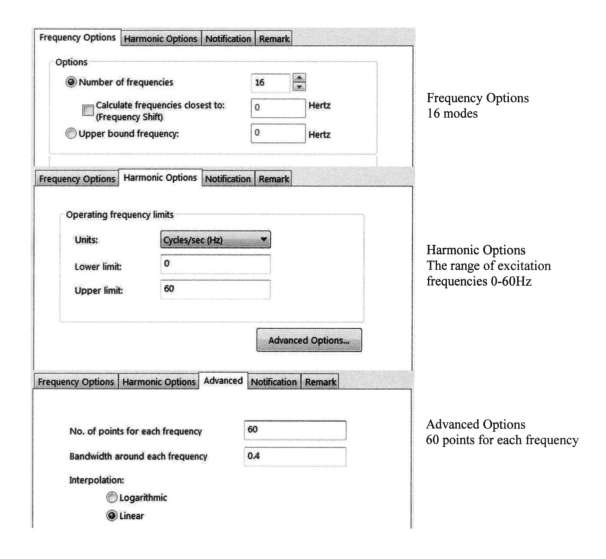

Frequency Options
16 modes

Harmonic Options
The range of excitation
frequencies 0-60Hz

Advanced Options
60 points for each frequency

Figure 19-13: Properties of Harmonic study *TEST 5 H.*

*16 modes are specified in the Frequency Options to satisfy the NAFEMS
TEST 5H requirements.*

Define the same restraints as in study *TEST 5*. Define a **Modal Damping** 2% for all modes; define load as shown in Figure 19-14.

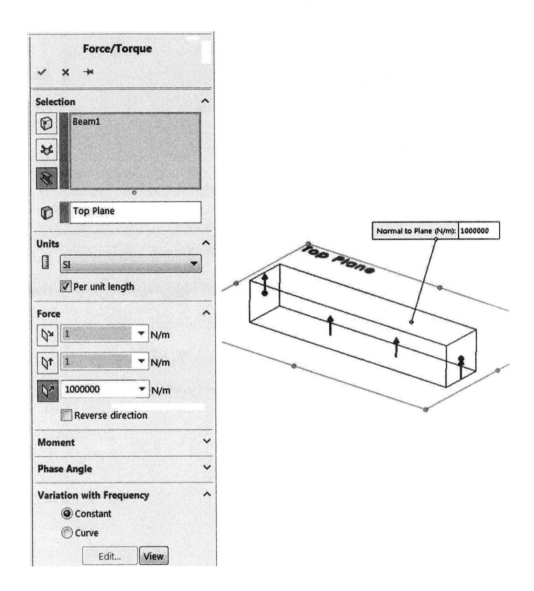

Figure 19-14: Load to be used in the Frequency Response study TEST 5H.

Considering the load is defined as 1e6N/m per unit length and the beam length is 10m, the total load is 1e7N.

Obtain the solution and notice that it completes in 121 steps. This is because we specified 60 points for each frequency in the **Advanced Options** of the study properties. Considering that there is only one frequency in the range of 0-60Hz, there are 60 points on each side of the natural frequency plus one point for the resonant frequency corresponding to the maximum response.

Construct the URES displacement plot for frequency step corresponding to the natural frequency 43.92Hz. Probe the plot in the midspan and select **Response Graph** as shown in Figure 19-15.

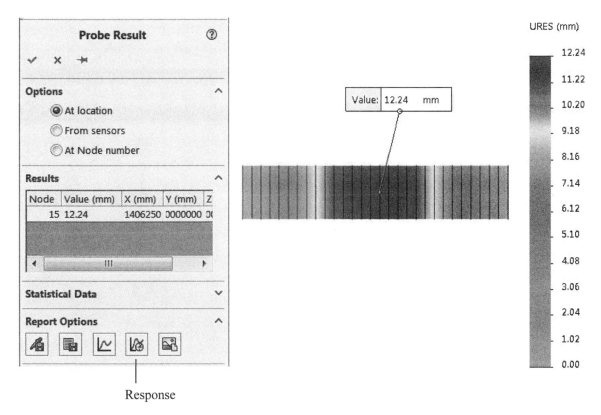

Response

Figure 19-15: Definition of URES displacement response graph.

Probe in the midspan of beam; select Response graph.

This way of creating a response graph doesn't require a sensor.

Repeat probing and creating a **Response graph** for **Upper bound axial and bending** stress plot. Both graphs are shown in Figure 19-16.

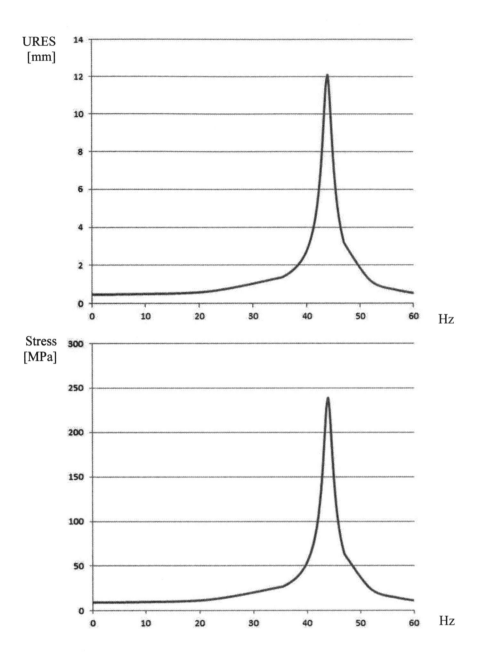

Figure 19-16: URES displacement (top) and Upper Bound Axial and Bending stress (bottom) in the mid-span of the beam as a function of the excitation frequency.

The peak corresponds to the natural frequency 43.92Hz.

Results of study TEST 5H report the maximum displacement 12.1mm and the maximum stress 239MPa while benchmark targets are 13.45mm and 241.9MPa.

NAFEMS TEST 5T

Deep Simply-Supported Beam; Transient Forced Vibration Response.

Linear solution with Modal Superposition

Nonlinear solution with Direct Integration

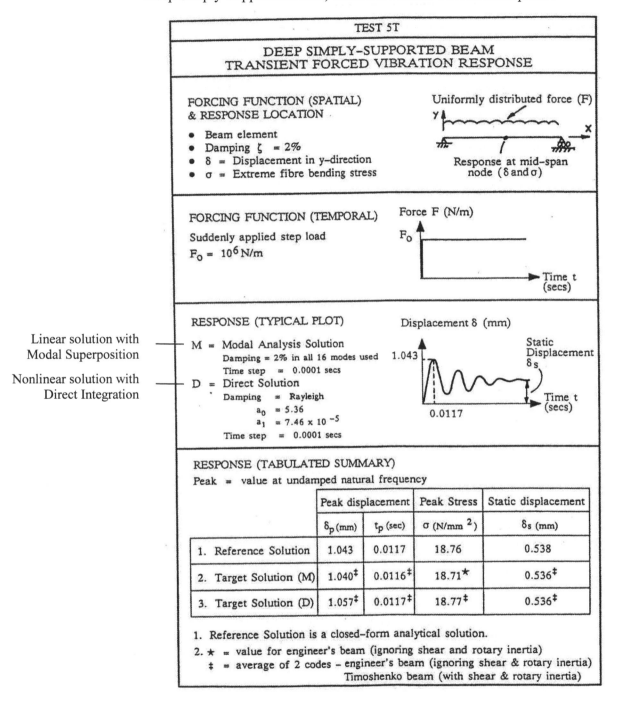

Figure 19-17: NAFEMS TEST 5T; beam elements are specified.

The model and material properties are the same as in NAFEMS TEST 5.

We'll use two solution methods as specified above: linear solution using the Modal Superposition and nonlinear solution using the Direct Integration.

We stay with the same model NAFEMS TEST 5; benchmark test NAFEMS TEST 5T will be conducted twice. First we obtain a linear solution with a **Modal Time History** study *TEST 5T LIN*; next a nonlinear solution with **Nonlinear Dynamic** study *TEST 5T NL*; both using configuration: *03 beams*.

The model is subjected to a suddenly applied load $F_0 = 10^6 N/m$; the load time history curve is shown in Figure 19-18.

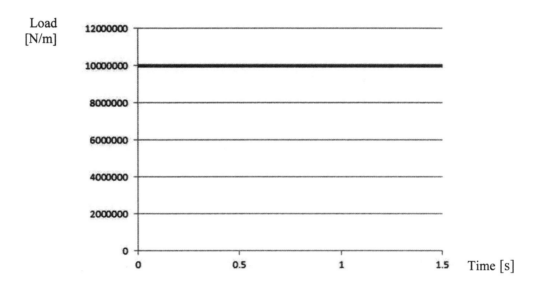

Figure 19-18: The load time history in NAFEMS TEST 5T.

The suddenly applied load of 100000N/m remains constant for the duration of the analysis of 1.5s.

The **Time Step** to be used is 0.0001s. The **Modal Damping** used in linear study is 2% in all 16 modes to be used for analysis. The **Rayleigh Damping**, to be used later in the nonlinear solution, is $\alpha = 5.36$, $\beta = 7.46e10^{-5}$.

The benchmark test target is the displacement URES = 1.043mm in the mid-span of beam at time t = 0.0117s; static displacement $\rho = 0.538$mm (Figure 19-19).

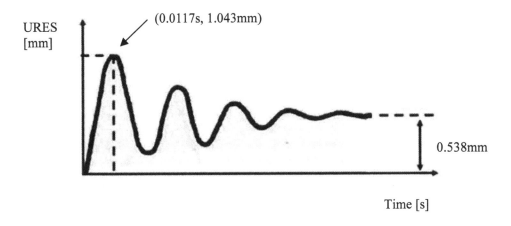

Figure 19-19: Targets in NAFEMS TEST 5T.

This illustration repeats information shown in Figure 19-17.

We start with the linear study. Create a **Modal Time History** study *TEST 5T LIN;* instruct **Simulation** to treat the solid body as a beam and define restraints identical to those in study *TEST 5.* Go through the **Edit Joints** dialog to recalculate the joints, and apply the load as shown in Figure 19-20.

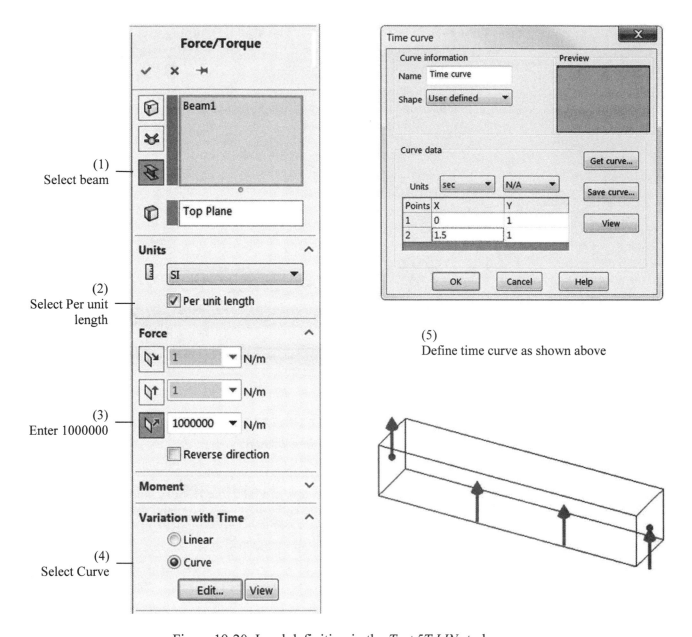

Figure 19-20: Load definition in the *Test 5T LIN* study.

The load is uniformly distributed over the beam. Considering the length of 10m, the total load is $10^7 N$. Model is shown in a wireframe display to show load which is applied to the center line corresponding to beam elements location.

The shock load duration of 1.5s is longer than the solution time 1s (Figure 19-21).

Define the **Modal Time History** study properties as shown in Figure 19-21.

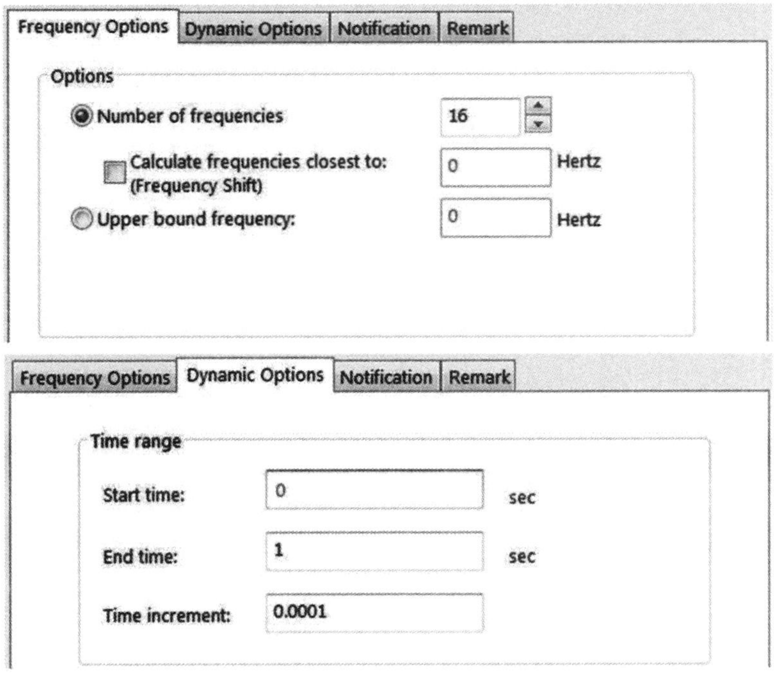

Figure 19-21: Properties of the Modal Time History (Time Response) study _TEST 5T LIN._

16 modes will be considered in the Modal Time History conducted in 10000 steps.

Define a **Modal Damping** of 2% for all modes and run the solution. Construct the URES displacement **Response Graph** as shown in Figure 19-22.

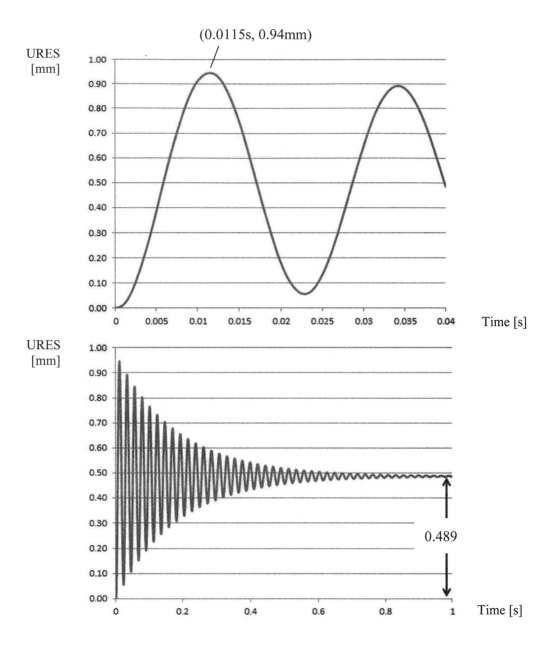

Figure 19-22: Results of linear study *TEST 5T LIN*. Time history of URES displacement in the mid-span of the beam during the first 0.04s (top) and during 1s (bottom).

After 1s the vibration almost disappears and the URES displacement at 1s is taken as the static displacement.

Results of study *TEST 5T LIN* report the maximum displacement 0.94mm at 0.0115s and the static displacement 0.489mm. Benchmark test targets are 1.043mm displacement at 0.0117s and static displacement 0.538mm.

We'll now repeat the benchmark test TEST 5T using the Direct Integration solution. Create a **Nonlinear Dynamic** study *TEST 5T NL* with properties as shown in Figure 19-23

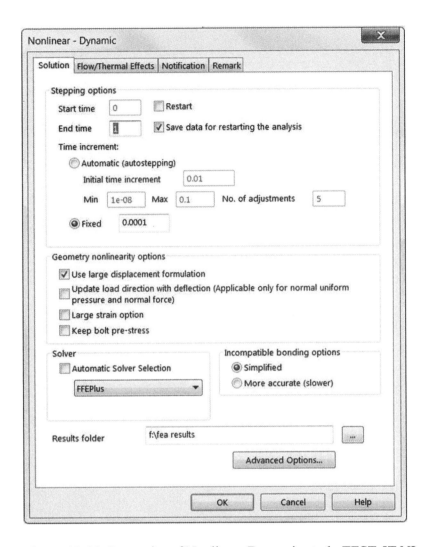

Figure 19-23: Properties of Nonlinear Dynamic study TEST 5T NL.

Duration of 1s combined with Fixed time step 0.0001s gives 10000 time steps during solution.

Define Rayleigh damping using constants: α =5.36, β =7.46e-5

Define restraints and load identical as in the linear study *TEST 5T LIN*.

Be prepared for a long solution time despite a very simple model. Specify a shorter solution time to solve this study faster.

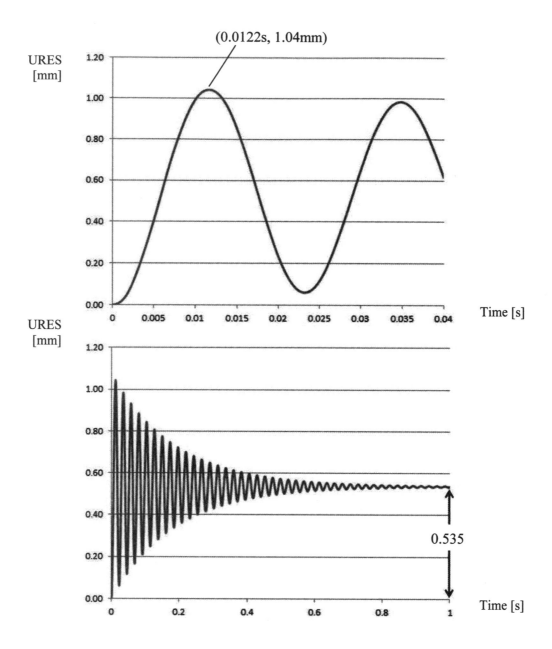

Figure 19-24: Results of linear study *TEST 5T NL*. Time history of URES displacement in the mid-span of the beam during the first 0.04s (top) and during 1s (bottom).

After 1s the vibration almost disappears and the URES displacements is taken as the static displacement.

Results of study *TEST 5T NL* report the maximum displacement 1.04mm at 0.0122s and the static displacement 0.535mm. Benchmark test targets are 1.043mm displacement at 0.0117s and static displacement 0.538mm.

NAFEMS TEST 13

Simply-Supported Thin Square Plate.

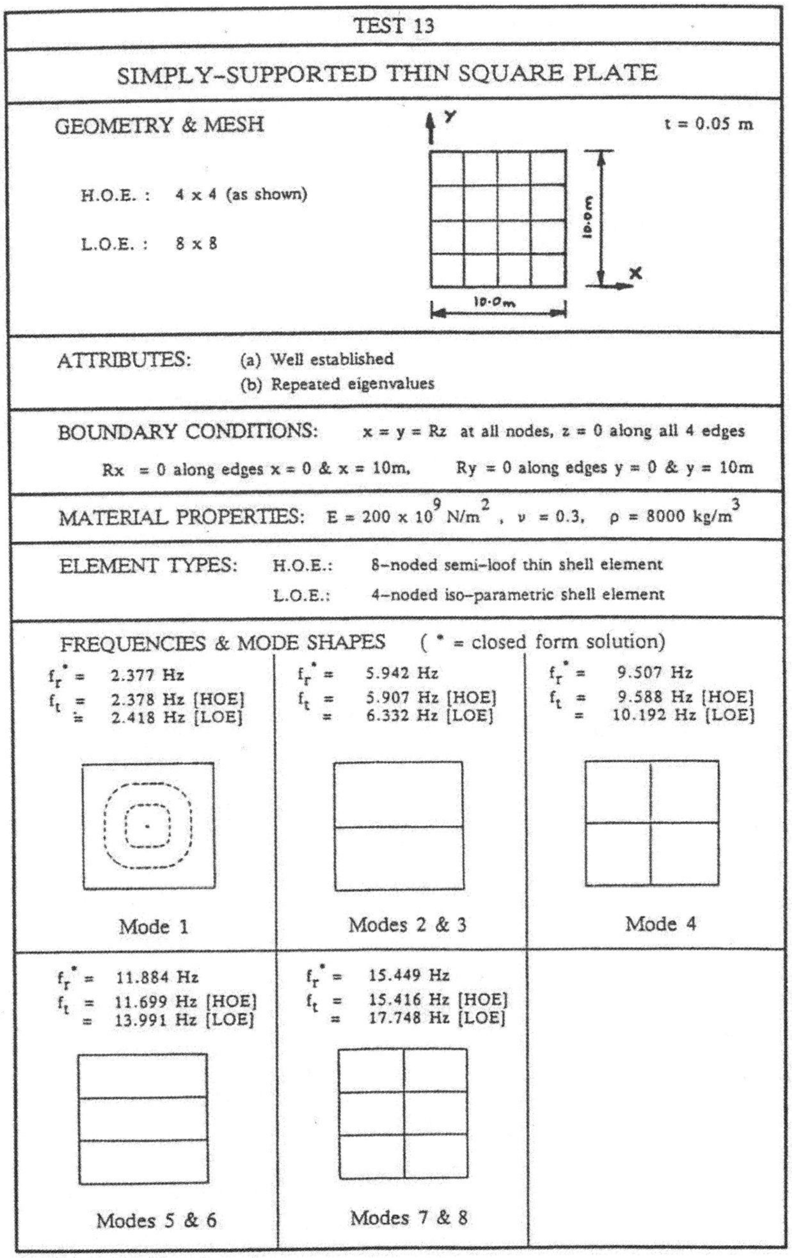

TEST 13

SIMPLY-SUPPORTED THIN SQUARE PLATE

GEOMETRY & MESH — t = 0.05 m

H.O.E. : 4 x 4 (as shown)

L.O.E. : 8 x 8

ATTRIBUTES: (a) Well established
(b) Repeated eigenvalues

BOUNDARY CONDITIONS: x = y = Rz at all nodes, z = 0 along all 4 edges
Rx = 0 along edges x = 0 & x = 10m, Ry = 0 along edges y = 0 & y = 10m

MATERIAL PROPERTIES: $E = 200 \times 10^9$ N/m^2, $\nu = 0.3$, $\rho = 8000$ kg/m^3

ELEMENT TYPES: H.O.E.: 8-noded semi-loof thin shell element
L.O.E.: 4-noded iso-parametric shell element

FREQUENCIES & MODE SHAPES (* = closed form solution)

Mode	f_r^*	f_t [HOE]	f_t [LOE]
Mode 1	2.377 Hz	2.378 Hz	2.418 Hz
Modes 2 & 3	5.942 Hz	5.907 Hz	6.332 Hz
Mode 4	9.507 Hz	9.588 Hz	10.192 Hz
Modes 5 & 6	11.884 Hz	11.699 Hz	13.991 Hz
Modes 7 & 8	15.449 Hz	15.416 Hz	17.748 Hz

Figure 19-25: NAFEMS TEST 13.

Review the boundary conditions.

Open part model NAFEMS TEST 13 and notice a reference point in the center; it is there to locate a sensor. Materials properties require a custom material, the same as those used in the previous tests.

Create a **Frequency** study titled *TEST 13*; specify eight modes in the study properties.

Define restraints to all four edges in the direction normal to the plate (Figure 19-26).

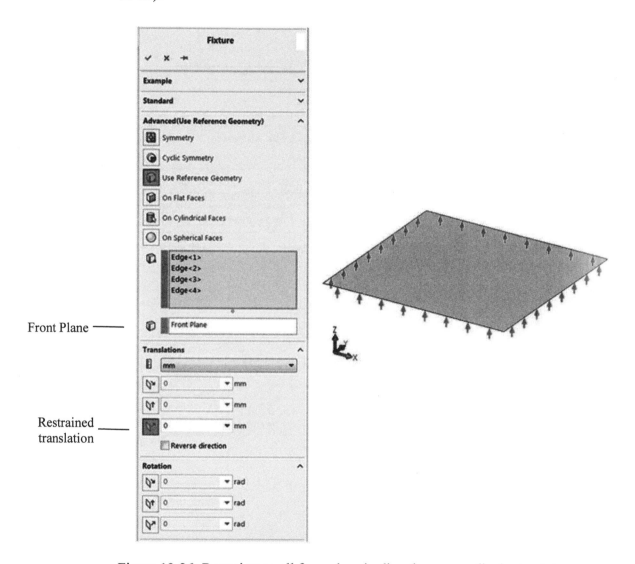

Front Plane ———

Restrained ____
translation

Figure 19-26: Restraints to all four edges in direction perpendicular to plate.

The Front Plane is used as reference geometry.

Reference point called Sensor is placed in the center of face; it will be used to locate Workflow Sensitive Sensor.

Define restraints to the two edges parallel to the X axis as shown in Figure 19-27.

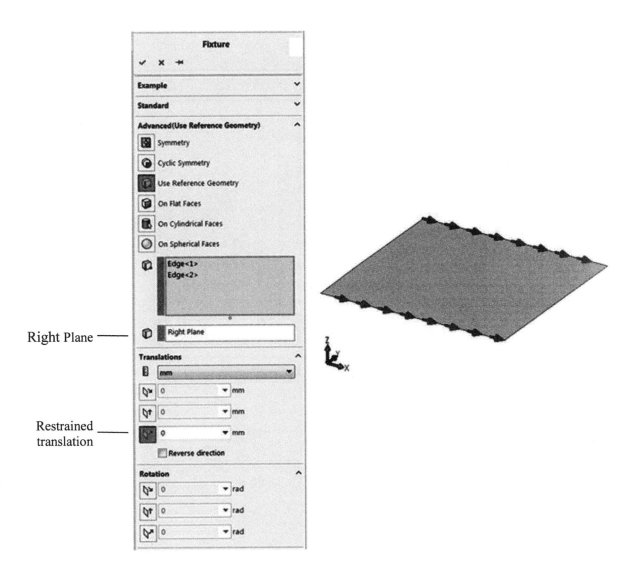

Figure 19-27: Restraints to the two edges parallel to the X axis.

The Right Plane is used as reference geometry.

Define restraints to the two edges parallel to the Y axis as shown in Figure 19-28.

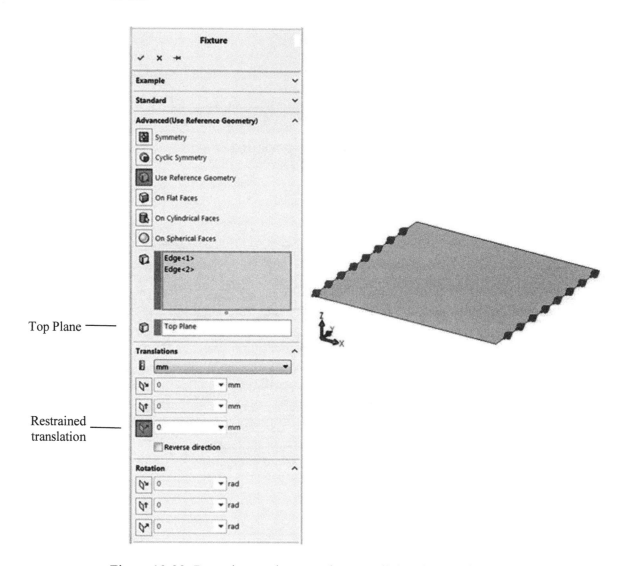

Figure 19-28: Restraints to the two edges parallel to the Y axis.

The Top Plane is used as reference geometry.

Restraints shown in Figure 19-26, Figure 19-27, and Figure 19-28 fully restrain the model. There is no need to apply x = y = Rz = 0 at all nodes as specified in the test.

Define the shell thickness as 50mm and mesh the model with the default element size.

The first two modes of vibration are shown in Figure 19-29.

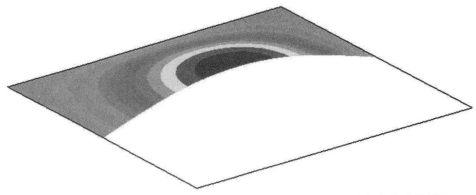

Mode 1: 2.374Hz

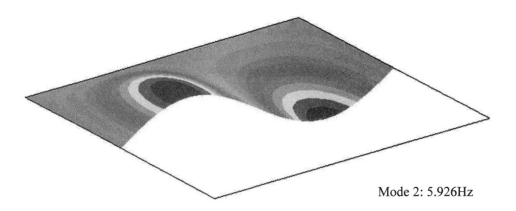

Mode 2: 5.926Hz

<u>Figure 19-29: The first two modes of vibration and their frequencies.</u>

Mode 1 is symmetric, mode 2 is anti-symmetric. Section Clipping is used to show the modal shapes more clearly.

The results of the modal analysis closely match the targets of NAFEMS TEST 13: 2.377Hz in the first mode and 5.942Hz in the second mode.

NAFEMS TEST 13R

Simply-Supported Thin Square Plate, Random Forced Vibration Response. This study uses the NAFEMS TEST 13 model.

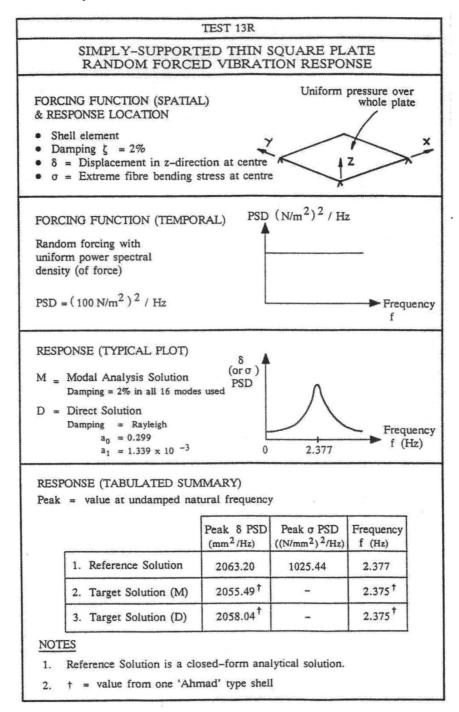

TEST 13R

| SIMPLY–SUPPORTED THIN SQUARE PLATE RANDOM FORCED VIBRATION RESPONSE |

FORCING FUNCTION (SPATIAL) & RESPONSE LOCATION

Uniform pressure over whole plate

- Shell element
- Damping ζ = 2%
- δ = Displacement in z–direction at centre
- σ = Extreme fibre bending stress at centre

FORCING FUNCTION (TEMPORAL)

Random forcing with uniform power spectral density (of force)

$$PSD = (100\ N/m^2)^2 / Hz$$

PSD $(N/m^2)^2$ / Hz

Frequency f

RESPONSE (TYPICAL PLOT)

M = Modal Analysis Solution
Damping = 2% in all 16 modes used

D = Direct Solution
Damping = Rayleigh
a_0 = 0.299
a_1 = 1.339 x 10^{-3}

δ (or σ) PSD

Frequency f (Hz)

0   2.377

RESPONSE (TABULATED SUMMARY)

Peak = value at undamped natural frequency

	Peak δ PSD (mm^2/Hz)	Peak σ PSD $((N/mm^2)^2/Hz)$	Frequency f (Hz)
1. Reference Solution	2063.20	1025.44	2.377
2. Target Solution (M)	2055.49[†]	–	2.375[†]
3. Target Solution (D)	2058.04[†]	–	2.375[†]

NOTES

1. Reference Solution is a closed–form analytical solution.

2. † = value from one 'Ahmad' type shell

Figure 19-30: NAFEMS TEST 13R.

The model geometry and restraints are identical to study TEST 13. The problem will be solved with the Modal Superposition Method.

Create a **Random** study titled *TEST 13R* and copy all restraints from the *TEST 13* study.

The force is a pressure; p = (100N/m²)²/Hz with a uniform Power Spectral Density which is a white noise. Define the load as shown in Figure 19-31. Remember that the pressure must be entered in units of $(N/m^2)^2/Hz$. Therefore, its numerical value is $p = (100N/m^2)^2/Hz = 100^2 \ (N/m^2)^2/Hz = 10000 \ (N/m^2)^2/Hz$. The use of incorrect units is a common source of error in Random Vibration analysis.

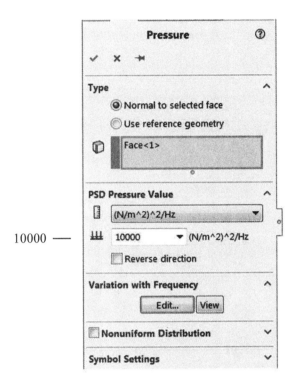
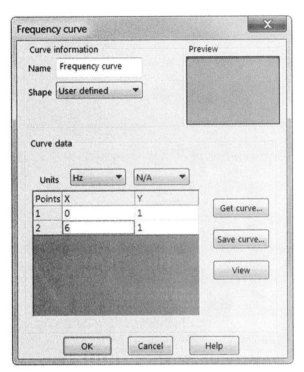

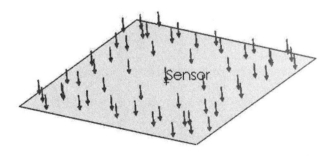

Figure 19-31: NAFEMS TEST 13R.

The frequency curve window defines a uniform PSD curve in the frequency range of 0-100Hz. Pay attention to units of pressure.

Define a **Modal Damping** of 2% for 16 modes in **Global Damping** definition window.

Define a **Random** study with properties as shown in Figure 19-32.

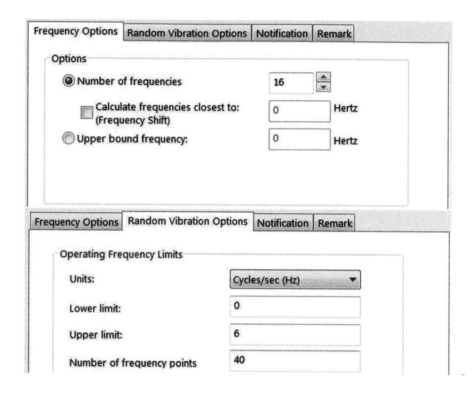

Figure 19-32: Properties of study *TEST 13R*.

The range 0-6Hz and 40 frequency points are specified in the Random Vibration Options.

Define the **Result Options** using the Sensor in the center of plate and obtain the solution.

Construct a **Response Graph** of the PSD displacement (Figure 19-33).

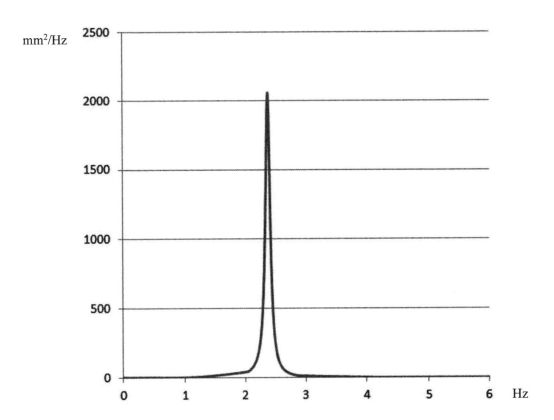

Figure 19-33: The PSD displacement response graph.

The first mode dominates the vibration response in the frequency range of 0-6Hz.

Review the graph and notice that the maximum PSD displacement closely matches the NAFEMS TEST 13R result of 2063mm^2/Hz.

To complete the analysis of Random Vibration problem, review RMS displacements plot shown in Figure 19-34.

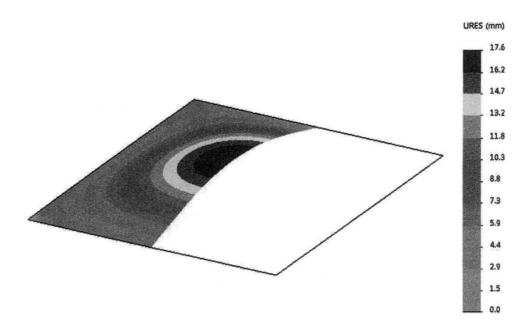

Figure 19-34: RMS value of displacement. One value of the RMS displacements is reported for the whole range of excitation 0-6Hz.

Section Clipping is used to present this plot.

20: Glossary of terms

Critical Damping

System damping above which oscillations are not possible.

Degrees of Freedom

Independent variables describing position (linear or angular) of a rigid body or of a node in a finite element mesh.

Discrete Vibrating System

System components responsible for inertial, damping and elastic properties are separated. It takes the form of an assembly of rigid bodies connected by springs and dampers.

Distributed Vibrating System

System components responsible for inertial, damping and elastic properties are not separated. It takes the form of an assembly of elastic bodies.

Dynamic Analysis

Analysis of motion of rigid bodies and flexible bodies.

Flexible body

Body that deforms under applied loads. Flexible body must have elastic properties defined.

Frequency Analysis

SOLIDWORKS Simulation term used to denote Modal Analysis.

Frequency Response Analysis

Analysis of vibration caused by loads defined as function of frequency rather than of time.

Impulse load

Load of a short duration defined as a function of time.

Harmonic Analysis

SOLIDWORKS Simulation term used to denote Frequency Response Analysis.

Kinematic Pair

Connection between two bodies that imposes constraints on relative movements of the bodies.

Linear Damping

Damping in linear motion.

Mechanism

System of rigid and/or elastic bodies connected by kinematic pairs, working together in a machine.

Modal Analysis

Finding modes of vibration of a vibrating system.

Modal Damping

% of critical damping present in a Single Degree of Freedom System.

Modal Superposition Method

Method that represents a vibrating system as a collection of Single Degree of Freedom Oscillators (SDOF); each SDOF represents the system vibrating in a given mode of vibration. The vibration response of the system is found as a superposition of vibration responses of those individual SDOFs.

Mode of Vibration

Certain combination of frequency and shape of vibration for which cancellation between inertial and elastic effects takes place.

Modal Time History

SOLIDWORKS Simulation term used to denote Time Response Analysis.

Random Vibration Analysis

Analysis of vibration due to stationary, random excitation given in the form of the Power Spectral Density.

Rayleigh Damping

Damping used in nonlinear vibration analysis; damping matrix is expressed as a linear combination of mass and stiffness matrices.

Response Spectrum Analysis

Type of vibration analysis used to analyze vibration caused by non-stationary excitation such as a seismic shock.

Rigid Body Mode

Mode of vibration of a structure without support or with a partial support. Program assigns 0Hz frequency to modes corresponding to rigid body displacements.

Rigid Body Motion

Translation and/or rotation of a body without deformation.

Single Degree of Freedom Oscillator

A mass (rigid body) connected to a base with a spring (and sometimes damper). It performs linear or angular oscillations and its motions is fully described by one unknown.

Structure

An elastic body capable of supporting loads. Structure is fully supported and it can't move without deforming. Movement of structure takes form of oscillations about the position of equilibrium.

Time Response Analysis

Analysis of vibration caused by loads defined as function time.

Vibration Analysis

Analysis of motion of a deformable (elastic) body or system about an equilibrium point.

Viscous Damping

Damping force is proportional to velocity.

21: References

For a review of the topics discussed in this book, readers are referred to the following literature listed here in the order of relevance:

1. Kurowski P. "Engineering Analysis with SOLIDWORKS Simulation"

2. Inman D., "Engineering Vibration" Prentice Hall

3. The Standard NAFEMS Benchmarks, NAFEMS;

4. Selected Benchmarks for Forced Vibration, NAFEMS;

5. How to Do Seismic Analysis Using Finite Elements NAFEMS;

6. How to undertake Finite Element Based Vibration Analysis NAFEMS;

7. Vibrationdata.com

8. Logan D. "A First Course in the Finite Element Method", Brooks/Cole

22: List of exercises

Chapter	Part	Assembly	Spreadsheet
Cover page	MESH2019 *		
1	TRUSS CLIP*	ELLIPTIC TRAMMEL** HELICOPTER PLANAR PRISMATIC REVOLUTE CYLINDRICAL SPHERICAL SCREW DISCRETE LINEAR* SWING ARM* DOUBLE PENDULUM WRECKING BALL** ROLLER****	
2	VALVE PLATE		
3	NOTCHED PLATE		
4	ROTOR COLUMN	COLUMN***	
5	U BRACKET		
6	BEAM DEMO*	SHAFT	
7	CAR		

*   Includes Simulation study ready to run
**   Includes Motion study ready to animate
***   For illustration only
**** Model used in SAE web seminar WB1401

Chapter	Part	Assembly	Spreadsheet
8	VASE		
9	BAJA FRAME		
10	FLAT*	PLIERS	
11		MDOF	MDOF.xlsx MDOF IMPULSE.xlsx
12	ELBOW PIPE	SHAKER TABLE***	
13		CENTRIFUGE	CENTRIFUGE.xlsx
14		CANTILEVER BEAM	
15		VIB ABSORBER	
16		HD HEAD	HD HEAD.xlsx
17		FRAME	FRAME.xlsx
18	PLANK		PLANK.xlsx PLANK RAYLEIGH.xlsx
19	NAFEMS TEST FV2 NAFEMS TEST 5 NAFEMS TEST 13		

*   Includes Simulation study ready to run
***   For illustration only

About Us

SDC Publications specializes in creating exceptional books that are designed to seamlessly integrate into courses or help the self learner master new skills. Our commitment to meeting our customer's needs and keeping our books priced affordably are just some of the reasons our books are being used by nearly 1,200 colleges and universities across the United States and Canada.

SDC Publications is a family owned and operated company that has been creating quality books since 1985. All of our books are proudly printed in the United States.

Our technology books are updated for every new software release so you are always up to date with the newest technology. Many of our books come with video enhancements to aid students and instructor resources to aid instructors.

Take a look at all the books we have to offer you by visiting SDCpublications.com.

NEVER STOP LEARNING

Keep Going

Take the skills you learned in this book to the next level or learn something brand new. SDC Publications offers books covering a wide range of topics designed for users of all levels and experience. As you continue to improve your skills, SDC Publications will be there to provide you the tools you need to keep learning. Visit SDCpublications.com to see all our most current books.

Why SDC Publications?

- Regular and timely updates
- Priced affordably
- Designed for users of all levels
- Written by professionals and educators
- We offer a variety of learning approaches

TOPICS

3D Animation
BIM
CAD
CAM
Engineering
Engineering Graphics
FEA / CAE
Interior Design
Programming

SOFTWARE

Adams
ANSYS
AutoCAD
AutoCAD Architecture
AutoCAD Civil 3D
Autodesk 3ds Max
Autodesk Inventor
Autodesk Maya
Autodesk Revit
CATIA
Creo Parametric
Creo Simulate
Draftsight
LabVIEW
MATLAB
NX
OnShape
SketchUp
SOLIDWORKS
SOLIDWORKS Simulation